OXFORD MATHEMATICAL MONOGRAPHS

Series Editors

E. M. FRIEDLANDER   I. G. MACDONALD   H. MCKEAN

R. PENROSE   J. T. STUART

OXFORD MATHEMATICAL MONOGRAPHS

A. Belleni-Morante: *Applied semigroups and evolution equations*

I. G. Macdonald: *Symmetric functions and Hall polynomials*

J. W. P. Hirschfeld: *Projective geometries over finite fields*

N. Woodhouse: *Geometric quantization*

A. M. Arthurs: *Complementary variational principles* Second edition

P. L. Bhatnagar: *Nonlinear waves in one-dimensional dispersive systems*

N. Aronszajn, T. M. Creese, and L. J. Lipkin: *Polyharmonic functions*

J. A. Goldstein: *Semigroups of linear operators*

M. Rosenblum and J. Rovnyak: *Hardy classes and operator theory*

J. W. P. Hirschfeld: *Finite projective spaces of three dimensions*

K. Iwasawa: *Local class field theory*

A. Pressley and G. Segal: *Loop groups*

J. C. Lennox and S. E. Stonehewer: *Subnormal groups of groups*

Wang Jianhua: *The theory of games*

S. Omatu and J. H. Seinfeld: *Distributed parameter system theory and applications*

The Theory of Games

WANG JIANHUA

Department of Applied Mathematics
Tsinghua University
Beijing, People's Republic of China

TSINGHUA UNIVERSITY PRESS · BEIJING
CLARENDON PRESS · OXFORD
1988

Oxford University Press, Walton Street, Oxford OX2 6DP

Oxford New York Toronto
Delhi Bombay Calcutta Madras Karachi
Petaling Jaya Singapore Hong Kong Tokyo
Nairobi Dar es Salaam Cape Town
Melbourne Auckland

and associated companies in
Berlin Ibadan

Oxford is a trade mark of Oxford University Press

Published in the United States
by Oxford University Press, New York

Originally published in Chinese
by Tsinghua University Press, 1986

This English edition © Tsinghua University Press,
and Oxford University Press 1988

British Library Cataloguing in Publication Data

Wang, Jianhua
The theory of games.
1. Game theory
I. Title
519.3
ISBN 0-19-853560-0

Library of Congress Cataloging in Publication Data

Wang, Jianhua.
The theory of games
(Oxford mathematical monographs)
Bibliography: p. 159
1. Game theory. I. Title. II. Series.
QA269.W35 1988 519.3 88-1876
ISBN 0-19-853560-0

Typeset and printed by The Universities Press (Belfast) Ltd

PREFACE

The purpose of this book is to present in a brief volume a systematic description of the main aspects of game theory, including not only the most fundamental concepts and properties, but also some of the results in the recent literature of the theory. In treating the subject matter I have tried to make it as concise as possible, and tedious enumerations have been avoided. For instance, regarding the optimal strategies of the matrix game and the continuous game on the unit square, only those properties which are most essential and most useful have been included. Others of lesser importance have been discarded without affecting the consistency of the material. Other subjects have been dealt with on similar principles.

Mathematical rigour has been kept throughout the entire book. Only a very few proofs have beem omitted to avoid leading the readers away from the main line of discussion. However, related literature is indicated as appropriate, and those readers who are interested in the problem under consideration can make further investigations without difficulty.

The theory of matrix games had its full development during the immediate years after J. von Neumann and O. Morgenstern published the classic work *Theory of games and economic behavior* in 1944. The first chapter of the present book is devoted chiefly to the relevant notions and properties of a matrix game. Two elementary proofs of the minimax theorem for the matrix game are given. One is von Neumann's proof based on the theorem of the supporting hyperplanes. The other is a very simple inductive proof. As to the relationship between matrix games and linear programming, it can be found in many books on game theory, linear programming, or operations research. I have therefore decided that this problem will just be touched upon very briefly in the present volume.

The theories of continuous games and non-cooperative games are introduced in Chapter 2 and Chapter 3 respectively.

In Chapter 4, my aim is to place stress on the several solution concepts of the *n*-person cooperative games, especially the concept of nucleolus, which is of great significance with reference to this class of game. Readers will be introduced to D. Schmeidler's results on this notion.

This book originated from a lecture course for graduate and undergraduate students at Tsinghua University in Beijing. The material can be taught in a one semester course totalling about 30 lecture hours.

Beijing   Wang Jianhua

CONTENTS

1

MATRIX GAMES

1.1 Introduction

Game theory is an important branch of operations research. A typical problem of the theory is this: two or more participants, called players, make decisions in a conflicting or competitive situation, each player aiming at reaching an outcome which is as advantageous to his own side as possible.

In order to have an intuitive understanding of a game, we first introduce some related basic ideas through a few simple examples.

Example 1.1. Matching pennies
Each of two participants, called player 1 and player 2, puts down a coin on the table without letting the other player see it. If the coins match, i.e. if both coins show heads or both show tails, player 1 wins the two coins. In other words, player 1 receives a payment of 1 from player 2. If they do not match, player 2 wins the two coins. That is to say, player 1 receives a payment of −1.

These outcomes can be listed in the following table:

		Player 2	
		1 (heads)	2 (tails)
Player 1	1 (heads)	1	−1
	2 (tails)	−1	1

We say that each player has two *strategies*. The first row represents the first strategy of player 1; the second row represents the second strategy of player 1. If player 1 chooses his strategy 1, it means that his coin shows heads up. Strategy 2 means tails up. Similarly, the first and second columns correspond respectively to the first and second strategies of player 2.

This gambling contest is a *game*. Briefly speaking, a game is a set of rules, in which the regulations of the entire procedure of competition (or contest, or struggle), including players, strategies, and the outcome after each play of the game is over, etc., are specifically described.

The entries in the above table form a *payoff matrix*. The *payoff* is a function of the strategies of the two players. If, for instance, player 1's coin shows heads up (strategy 1) and player 2's coin also shows heads up (strategy 1), then the element 1 in the first row and first column denotes the amount which player 1 receives from player 2. Again, if player 1 chooses strategy 2 (tails) and player 2 chooses strategy 1 (heads), then the element -1 in the second row and first column is the payment that player 1 receives. In this case, the payment player 1 receives is a negative number. This means that player 1 loses one unit. In other words, player 1 pays one unit to player 2.

Example 1.2. 'Stone–paper–scissors'
All children play this game. Scissors defeats paper, paper defeats stone, and stone in turn defeats scissors. There are two players: 1 and 2. Each player has three strategies. Let strategies 1, 2, 3 represent stone, paper, scissors, respectively. Suppose that the winner wins one unit from the loser, then the payoff matrix is

		Player 2		
		1	2	3
	1	0	-1	1
Player 1	2	1	0	-1
	3	-1	1	0

Example 1.3. Player 1 chooses a number from among four numbers $p = 0, 1, 2, 3$ and player 2 without knowing player 1's choice chooses a number from among three numbers $q = 0, 1, 2$. The payoff to player 1 (i.e. the amount player 2 pays to player 1) is determined by the function

$$P = p(q - p) + q(q + p)$$
or
$$P = q^2 - p^2 + 2pq.$$

This is a two-person game.. Player 1 has four strategies, while player 2 has three strategies. The payoff matrix is easily evaluated:

	p \ q	Player 2		
		0	1	2
	0	0	1	4
Player 1	1	-1	2	7
	2	-4	1	8
	3	-9	-2	7

In each of the above examples, there are two players, namely player 1 and player 2, and a payoff matrix. Each player has several strategies. The strategies of player 1 are represented by the row numbers of the payoff matrix and those of player 2 by the column numbers. Player 1 chooses a strategy from his strategy set and player 2 independently chooses a strategy from his strategy set. After the two choices have been made, player 2 pays an amount to player 1 as the outcome of this particular play of the game. The amount is shown in the payoff matrix. If the payoff is positive, player 1 receives a positive amount from player 2, i.e. player 1 wins an amount from player 2. If the payoff is negative, player 1 receives a negative amount from player 2, i.e. player 1 loses an amount to player 2.

In each of these games, the gain of player 1 equals the loss of player 2. What player 1 wins is just what player 2 loses, and *vice versa*. Such a game is called a *zero-sum* game. Zero-sum games with two players are the simplest of all games and are called *zero-sum two-person games*.

If the number of players is *n*, the game is called an *n*-person game. *N*-person games can be divided into non-cooperative games $(n \geq 2)$ and cooperative games $(n \geq 2)$. We shall first introduce the theory of zero-sum two-person games. In this chapter the discussion will be restricted to matrix games, namely the zero-sum games in which both players 1 and 2 have a finite number of strategies.

1.2   Matrix games

Suppose player 1 has *m* strategies $i = 1, \ldots, m$; player 2 has *n* strategies $j = 1, \ldots, n$. Let a_{ij} be the payoff which player 1 gains from player 2 if player 1 chooses strategy *i* and player 2 chooses strategy *j*. Then the payoff matrix is

$$A = (a_{ij}) = \begin{bmatrix} a_{11} & a_{12} & \cdots & a_{1n} \\ a_{21} & a_{22} & \cdots & a_{2n} \\ \multicolumn{4}{c}{\cdots\cdots\cdots\cdots\cdots\cdots} \\ a_{m1} & a_{m2} & \cdots & a_{mn} \end{bmatrix}. \tag{1.1}$$

The game is completely determined by the above matrix. It is, therefore, called a *matrix game*.

In this game, player 1 wishes to gain as large a payoff a_{ij} as possible, while player 2 will do his best to reach as small a value of a_{ij} as possible. The interests of the two players are completely conflicting.

If player 1 chooses his first strategy, $i = 1$, he can be sure to obtain at least the payoff

$$\min_{1 \leq j \leq n} a_{1j}.$$

In general, if player 1 adopts the strategy i, his payoff is at least

$$\min_{1 \leq j \leq n} a_{ij}, \tag{1.2}$$

which is the minimum of the ith-row elements in the payoff matrix. Since player 1 wishes to maximize his payoff, he can choose i so as to make (1.2) as large as possible. That is to say, player 1 can choose i so as to receive a payoff not less than

$$\max_{1 \leq i \leq m} \min_{1 \leq j \leq n} a_{ij}. \tag{1.3}$$

Similarly, if player 2 chooses his strategy $j = 1$, he will lose at most

$$\max_{1 \leq i \leq m} a_{i1}.$$

In general, if player 2 chooses the strategy j, his maximum loss is

$$\max_{1 \leq i \leq m} a_{ij}, \tag{1.4}$$

which is the maximum of the jth-column elements in the payoff matrix. Since player 2 wishes to minimize the payoff, he will try to choose j so as to make (1.4) a minimum. That is to say, player 2 can choose j so as to make his loss not greater than

$$\min_{1 \leq j \leq n} \max_{1 \leq i \leq m} a_{ij}. \tag{1.5}$$

In other words, if player 2 makes his best choice, the payoff which player 1 receives cannot be greater than (1.5).

We have seen that player 1 can choose i to ensure a payoff which is at least $\max_{1 \leq i \leq m} \min_{1 \leq j \leq n} a_{ij}$, while player 2 can choose j to make player 1 get at most $\min_{1 \leq j \leq n} \max_{1 \leq i \leq m} a_{ij}$: is there any relationship between these two values?

Let us examine the three examples in the last section. In Example 1.1,

$$\max_{1 \leq i \leq m} \min_{1 \leq j \leq n} a_{ij} = \max(-1, -1) = -1,$$

$$\min_{1 \leq j \leq n} \max_{1 \leq i \leq m} a_{ij} = \min(1, 1) = 1.$$

We have

$$\max_{1 \leq i \leq m} \min_{1 \leq j \leq n} a_{ij} < \min_{1 \leq j \leq n} \max_{1 \leq i \leq m} a_{ij}.$$

In Example 1.2,

$$\max_{1 \leq i \leq m} \min_{1 \leq j \leq n} a_{ij} = -1 < 1 = \min_{1 \leq j \leq n} \max_{1 \leq i \leq m} a_{ij}.$$

In Example 1.3,

$$\max_{1\le i\le m}\min_{1\le j\le n} a_{ij} = \max(0, -1, -4, -9) = 0,$$

$$\min_{1\le j\le n}\max_{1\le i\le m} a_{ij} = \min(0, 2, 8) = 0.$$

The two values are equal.

We see that $\max_{1\le i\le m}\min_{1\le j\le n} a_{ij}$ and $\min_{1\le j\le n}\max_{1\le i\le m} a_{ij}$ may or may not be equal. In general, we have the following inequality:

$$\max_{1\le i\le m}\min_{1\le j\le n} a_{ij} \le \min_{1\le j\le n}\max_{1\le i\le m} a_{ij}. \qquad (1.6)$$

The proof of (1.6) is as follows. For every i we have

$$\min_{1\le j\le n} a_{ij} \le a_{ij}, \qquad j = 1, \ldots, n;$$

and for every j we have

$$a_{ij} \le \max_{1\le i\le m} a_{ij}, \qquad i = 1, \ldots, m.$$

Hence the inequality

$$\min_{1\le j\le n} a_{ij} \le \max_{1\le i\le m} a_{ij}$$

holds for all i and all j. Since the left-hand side of the last inequality is independent of j, taking the minimum with respect to j on both sides we get

$$\min_{1\le j\le n} a_{ij} \le \min_{1\le j\le n}\max_{1\le i\le m} a_{ij}, \qquad i = 1, \ldots, m.$$

Therefore,

$$\max_{1\le i\le m}\min_{1\le j\le n} a_{ij} \le \min_{1\le j\le n}\max_{1\le i\le m} a_{ij}$$

and the proof is completed.

1.3 Saddle points

If the elements of the payoff matrix (a_{ij}) of a matrix game satisfy

$$\max_{1\le i\le m}\min_{1\le j\le n} a_{ij} = v = \min_{1\le j\le n}\max_{1\le i\le m} a_{ij}, \qquad (1.7)$$

the quantity v is called the *value* of the game. v is the common value of (1.3) and (1.5)..

The value of the game of Example 1.3 in the last section is $v = 0$.

If the eqn (1.7) holds, there exist an i^* and a j^* such that

$$\min_{1 \leq j \leq n} a_{i^*j} = \max_{1 \leq i \leq m} \min_{1 \leq j \leq n} a_{ij} = v$$

and

$$\max_{1 \leq i \leq m} a_{ij^*} = \min_{1 \leq j \leq n} \max_{1 \leq i \leq m} a_{ij} = v.$$

Hence

$$\min_{1 \leq j \leq n} a_{i^*j} = \max_{1 \leq i \leq m} a_{ij^*}.$$

But

$$\min_{1 \leq j \leq n} a_{i^*j} \leq a_{i^*j^*} \leq \max_{1 \leq i \leq m} a_{ij^*}.$$

Thus

$$\max_{1 \leq i \leq m} a_{ij^*} = a_{i^*j^*} = v = \min_{1 \leq j \leq n} a_{i^*j}.$$

Therefore, for all i and all j we have

$$a_{ij^*} \leq a_{i^*j^*} = v \leq a_{i^*j}. \tag{1.8}$$

That is to say, if player 1 chooses the strategy i^*, then the payoff cannot be less than v if player 2 departs from the strategy j^*; if player 2 chooses the strategy j^*, then the payoff cannot exceed v if player 1 departs from the strategy i^*.

We call i^* and j^* *optimal strategies* of players 1 and 2 respectively. (i^*, j^*) is a *saddle point* of the game. We also say that $i = i^*$, $j = j^*$ is a *solution* of the game.

Relation (1.8) shows that the payoff at the saddle point (i^*, j^*) is the value of the game. As long as player 1 sticks to his optimal strategy i^*, he can hope to increase his payoff if player 2 departs from his optimal strategy j^*. Similarly, if player 2 sticks to his optimal strategy j^*, player 1's payoff may decrease if he departs from his optimal strategy i^*.

It is easy to prove that (1.8) is also a sufficient condition for (1.7) (see Section 1.7). Thus if the game has a saddle point (i^*, j^*), eqn (1.7) holds and $a_{i^*j^*} = v$.

A matrix game may have more than one saddle point. However, the payoffs at different saddle points are all equal, the common value being the value of the game.

Example 1.4. Consider the matrix game with the payoff matrix

$$A = \begin{bmatrix} 1 & -1 & 0 & 3 \\ -2 & -3 & -1 & -3 \\ 2 & 2 & 3 & 4 \end{bmatrix}.$$

It is easily verified that (3.1) and (3.2) are both saddle points and

$$a_{31} = a_{32} = v = 2.$$

If the matrix game (a_{ij}) has a saddle point (i^*, j^*), it can very easily be detected. By the definition of a saddle point (1.8), the value $a_{i^*j^*}$ is an element in the payoff matrix (a_{ij}) which is at the same time the minimum of its row and the maximum of its column.

In Example 1.3, $i^* = 1$, $j^* = 1$ is a saddle point of the game. $a_{11} = 0$ is the smallest element in the first row and at the same time the largest element in the first column. In Example 1.4, $a_{31} = a_{32} = 2$ are two smallest elements in the third row, and at the same time the largest elements in the first and second columns respectively.

When the number of saddle points of a matrix game exceeds one, we have the following theorem.

Theorem 1.1. *Let (i^*, j^*) and (i^0, j^0) be saddle points of a matrix game (a_{ij}). Then (i^*, j^0) and (i^0, j^*) are its saddle points, and the values at all saddle points are equal, i.e.*

$$a_{i^*j^*} = a_{i^0j^0} = a_{i^*j^0} = a_{i^0j^*}. \tag{1.9}$$

Proof. Since (i^*, j^*) is a saddle point,

$$a_{ij^*} \leqslant a_{i^*j^*} \leqslant a_{i^*j} \tag{1.10}$$

for all i and all j. Since (i^0, j^0) is a saddle point,

$$a_{ij^0} \leqslant a_{i^0j^0} \leqslant a_{i^0j} \tag{1.11}$$

for all i and all j. From (1.10) and (1.11) we have

$$a_{i^*j^*} \leqslant a_{i^*j^0} \leqslant a_{i^0j^0} \leqslant a_{i^0j^*} \leqslant a_{i^*j^*}.$$

This proves (1.9).

By (1.9) and (1.10), (1.11),

$$a_{ij^0} \leqslant a_{i^*j^0} \leqslant a_{i^*j}$$

for all i and all j. Hence (i^*, j^0) is a saddle point.

The fact that (i^0, j^*) is a saddle point can be proved in a similar way. $\square$

We see from this theorem that a matrix game with saddle points has the following properties: the exchangeability or rectangular property of saddle points, and the equality of the values at all saddle points.

1.4 Mixed strategies

When a matrix game has no saddle point, i.e. if

$$\max_{1 \leqslant i \leqslant m} \min_{1 \leqslant j \leqslant n} a_{ij} < \min_{1 \leqslant j \leqslant n} \max_{1 \leqslant i \leqslant m} a_{ij}, \tag{1.12}$$

we cannot solve the game in the sense of the last section. For instance, the payoff matrix of the 'stone–paper–scissors' game in Section 1.1 is

$$A = \begin{bmatrix} 0 & -1 & 1 \\ 1 & 0 & -1 \\ -1 & 1 & 0 \end{bmatrix}.$$

We have seen that

$$\max_{1 \leqslant i \leqslant m} \min_{1 \leqslant j \leqslant n} a_{ij} = -1 < 1 = \min_{1 \leqslant j \leqslant n} \max_{1 \leqslant i \leqslant m} a_{ij}.$$

Player 1 can be sure to gain at least -1, player 2 can guarantee that his loss is at most 1. In this situation, player 1 will try to gain a payoff greater than -1, player 2 will try to make the payoff (to player 1) less than 1. For these purposes, each player will make efforts to prevent his opponent from finding out his actual choice of strategy. To accomplish this, player 1 can make use of some chance device to determine which strategy he is going to choose; similarly, player 2 will also decide his choice of strategy by some chance method. This is the mixed strategy to be introduced in this section.

Let the payoff matrix of a matrix game be $A = (a_{ij})$, where $i = 1, \ldots, m; j = 1, \ldots, n$. A *mixed strategy* of player 1 is a set of numbers $x_i \geqslant 0$, $i = 1, \ldots, m$, satisfying

$$\sum_{i=1}^{m} x_i = 1.$$

A mixed strategy of player 2 is a set of numbers $y_j \geqslant 0$, $j = 1, \ldots, n$, satisfying

$$\sum_{j=1}^{n} y_j = 1.$$

In contrast to mixed strategies, the strategies in the preceding sections are called *pure strategies*. The pure strategy $i = i'$ is in reality a special mixed strategy: $x_{i'} = 1$; $x_i = 0$ if $i \neq i'$.

Let $X = (x_1, \ldots, x_m)$ and $Y = (y_1, \ldots, y_n)$ be respective mixed strategies of players 1 and 2. Player 1 chooses his strategy i with probability x_i; player 2 chooses his strategy j with probability y_j. Hence, $x_i y_j$ is the probability that player 1 chooses strategy i and player 2 chooses strategy j

with payoff a_{ij}. Multiplying every payoff a_{ij} by the corresponding probability $x_i y_j$ and summing for all i and all j, we obtain the *expected payoff* of player 1:

$$\sum_{i=1}^{m} \sum_{j=1}^{n} a_{ij} x_i y_j. \tag{1.13}$$

Player 1 wishes to maximize this expected payoff, while player 2 wants to minimize it.

Let S_m be the set of all $X = (x_1, \ldots, x_m)$ satisfying the conditions

$$x_i \geqslant 0, \qquad i = 1, \ldots, m, \qquad \sum_{i=1}^{m} x_i = 1.$$

If player 1 uses the (mixed) strategy $X \in S_m$, then his expected payoff is at least

$$\min_{Y \in S_n} \sum_{i=1}^{m} \sum_{j=1}^{n} a_{ij} x_i y_j, \tag{1.14}$$

where S_n is the set of all $Y = (y_1, \ldots, y_n)$ satisfying the conditions

$$y_j \geqslant 0, \qquad j = 1, \ldots, n, \qquad \sum_{j=1}^{n} y_j = 1.$$

Player 1 can choose $X \in S_m$ so as to make (1.14) a maximum, i.e. he can be sure of an expected payoff not less than

$$v_1 = \max_{X \in S_m} \min_{Y \in S_n} \sum_{i=1}^{m} \sum_{j=1}^{n} a_{ij} x_i y_j. \tag{1.15}$$

If player 2 chooses the strategy $Y \in S_n$, then the expected payoff of player 1 is at most

$$\max_{X \in S_m} \sum_{i=1}^{m} \sum_{j=1}^{n} a_{ij} x_i y_j. \tag{1.16}$$

Player 2 can choose $Y \in S_n$ so as to make (1.16) a minimum, i.e. he can prevent player 1 from gaining an expected payoff greater than

$$v_2 = \min_{Y \in S_n} \max_{X \in S_m} \sum_{i=1}^{m} \sum_{j=1}^{n} a_{ij} x_i y_j. \tag{1.17}$$

We have Theorem 1.2 as follows.

Theorem 1.2

$$\max_{X \in S_m} \min_{Y \in S_n} \sum_{i=1}^{m} \sum_{j=1}^{n} a_{ij} x_i y_j \leqslant \min_{Y \in S_n} \max_{X \in S_m} \sum_{i=1}^{m} \sum_{j=1}^{n} a_{ij} x_i y_j. \tag{1.18}$$

Proof. For all $X \in S_m$ and all $Y \in S_n$, we have

$$\min_{Y \in S_n} \sum_{i=1}^{m} \sum_{j=1}^{n} a_{ij} x_i y_j \leqslant \sum_{i=1}^{m} \sum_{j=1}^{n} a_{ij} x_i y_j.$$

Taking the maximum for all $X \in S_m$ on both sides of the inequality, we obtain

$$\max_{X \in S_m} \min_{Y \in S_n} \sum_{i=1}^{m} \sum_{j=1}^{n} a_{ij} x_i y_j \leqslant \max_{X \in S_m} \sum_{i=1}^{m} \sum_{j=1}^{n} a_{ij} x_i y_j.$$

This inequality holds for all $Y \in S_n$. Therefore,

$$\max_{X \in S_m} \min_{Y \in S_n} \sum_{i=1}^{m} \sum_{j=1}^{n} a_{ij} x_i y_j \leqslant \min_{Y \in S_n} \max_{X \in S_m} \sum_{i=1}^{m} \sum_{j=1}^{n} a_{ij} x_i y_j. \qquad \square$$

J. von Neumann first proved that for all matrix games the two quantities (1.15) and (1.17) are equal. This result is the well-known *fundamental theorem of the theory of games*, or the *minimax theorem*.

1.5   The minimax theorem

Many proofs of the minimax theorem have been published in the game theory literature. We present here von Neumann's proof as is given in von Neumann and Morgenstern (1944).

Definition 1.1. Let

$$a^{(1)} = (a_{11}, a_{21}, \ldots, a_{m1}),$$
$$a^{(2)} = (a_{12}, a_{22}, \ldots, a_{m2}),$$
$$\ldots\ldots\ldots\ldots\ldots\ldots\ldots\ldots\ldots\ldots$$
$$a^{(n)} = (a_{1n}, a_{2n}, \ldots, a_{mn})$$

be n points in the m-dimensional Euclidean space. If the point

$$a = (a_1, a_2, \ldots, a_m)$$

can be expressed as a convex linear combination of the n points $a^{(1)}, \ldots, a^{(n)}$, i.e. if there exist

$$t_k \geqslant 0, \qquad k = 1, \ldots, n, \qquad \sum_{k=1}^{n} t_k = 1$$

such that

$$a = t_1 a^{(1)} + t_2 a^{(2)} + \cdots + t_n a^{(n)},$$

we say that the point a belongs to the *convex hull H* of $a^{(1)}, \ldots, a^{(n)}$.

H is indeed a convex set. This can easily be verified by showing that every convex linear combination of two arbitrary points of H also belongs to H.

In order to prove the minimax theorem, we need the following two lemmas.

Lemma 1. *Let H be the convex hull of*

$$a^{(1)} = (a_{11}, a_{21}, \ldots, a_{m1}),$$
$$a^{(2)} = (a_{12}, a_{22}, \ldots, a_{m2}),$$
$$\ldots\ldots\ldots\ldots\ldots\ldots\ldots\ldots\ldots\ldots$$
$$a^{(n)} = (a_{1n}, a_{2n}, \ldots, a_{mn}).$$

Suppose the origin 0 does not belong to H. Then there exist m real numbers $s_1, \ldots, s_m$ such that for every point

$$a = (a_1, a_2, \ldots, a_m)$$

of H we have

$$s_1 a_1 + s_2 a_2 + \cdots + s_m a_m > 0.$$

Proof. Since $0 \notin H$, there exists a point

$$s = (s_1, \ldots, s_m) \in H,$$

which is different from 0, such that the distance $|s|$ from s to 0 is smallest. This is equivalent to the statement that

$$s_1^2 + \cdots + s_m^2 > 0$$

is smallest.

Now let $a = (a_1, \ldots, a_m)$ be an arbitrary point in H. Then

$$\lambda a + (1 - \lambda)s \in H, \qquad 0 \leqslant \lambda \leqslant 1,$$

and

$$|\lambda a + (1 - \lambda)s|^2 \geqslant |s|^2,$$

or

$$\sum_{i=1}^{m} [\lambda a_i + (1 - \lambda)s_i]^2 = \sum_{i=1}^{m} [\lambda(a_i - s_i) + s_i]^2$$

$$= \lambda^2 \sum_{i=1}^{m} (a_i - s_i)^2 + 2\lambda \sum_{i=1}^{m} (a_i - s_i)s_i + \sum_{i=1}^{m} s_i^2$$

$$\geqslant \sum_{i=1}^{m} s_i^2.$$

If $\lambda \neq 0$,

$$\lambda \sum_{i=1}^{m} (a_i - s_i)^2 + 2 \sum_{i=1}^{m} (a_i s_i - s_i^2) \geq 0.$$

Let $\lambda \to 0$; we obtain

$$\sum_{i=1}^{m} a_i s_i \geq \sum_{i=1}^{m} s_i^2 > 0. \qquad \square$$

This lemma is usually referred to as the theorem of the supporting hyperplanes. It states that if the origin 0 does not belong to the convex hull H of the n points $a^{(1)}, \ldots, a^{(n)}$, then there exists a supporting hyperplane p passing through 0 such that H lies entirely in one side of p, i.e. in one of the two half-spaces formed by p.

Lemma 2. *Let $A = (a_{ij})$ be an arbitrary $m \times n$ matrix. Then either*
  (1) *there exist numbers $y_1, \ldots, y_n$ with*

$$y_j \geq 0, \qquad j = 1, \ldots, n, \qquad \sum_{j=1}^{n} y_j = 1,$$

$$\sum_{j=1}^{n} a_{ij} y_j = a_{i1} y_1 + a_{i2} y_2 + \cdots + a_{in} y_n \leq 0, \qquad i = 1, \ldots, m;$$

or
  (2) *there exist numbers $x_1, \ldots, x_m$ with*

$$x_i \geq 0, \qquad i = 1, \ldots, m, \qquad \sum_{i=1}^{m} x_i = 1,$$

$$\sum_{i=1}^{m} a_{ij} x_i = a_{1j} x_1 + a_{2j} x_2 + \cdots + a_{mj} x_m > 0, \qquad j = 1, \ldots, n.$$

Proof. Let H be the convex hull of the $n + m$ points

$$a^{(1)} = (a_{11}, a_{21}, \ldots, a_{m1}),$$
$$a^{(2)} = (a_{12}, a_{22}, \ldots, a_{m2}),$$
$$\cdots\cdots\cdots\cdots\cdots\cdots\cdots\cdots\cdots$$
$$a^{(n)} = (a_{1n}, a_{2n}, \ldots, a_{mn}),$$
$$e^{(1)} = (1, 0, \ldots, 0),$$
$$e^{(2)} = (0, 1, \ldots, 0),$$
$$\cdots\cdots\cdots\cdots\cdots\cdots\cdots\cdots\cdots$$
$$e^{(m)} = (0, 0, \ldots, 1).$$

We distinguish two cases:

(1) $0 \in H$. Then 0 can be expressed as a convex linear combination of the above $n + m$ points. That is to say, there exist numbers

$$t_1, t_2, \ldots, t_{n+m} \geq 0, \qquad \sum_{j=1}^{n+m} t_j = 1,$$

such that

$$t_1 a^{(1)} + t_2 a^{(2)} + \cdots + t_n a^{(n)} + t_{n+1} e^{(1)} + t_{n+2} e^{(2)} + \cdots + t_{n+m} e^{(m)} = 0.$$

Expressed in terms of the components, the ith equation (there is a total of m equations) is

$$t_1 a_{i1} + t_2 a_{i2} + \cdots + t_n a_{in} + t_{n+i} \cdot 1 = 0.$$

Hence

$$t_1 a_{i1} + t_2 a_{i2} + \cdots + t_n a_{in} = -t_{n+i} \leq 0, \qquad i = 1, \ldots, m. \qquad (1.19)$$

It follows that

$$t_1 + t_2 + \cdots + t_n > 0,$$

for otherwise we would have

$$t_1 = t_2 = \cdots = t_n = 0 = t_{n+1} = \cdots = t_{n+m},$$

which contradicts

$$\sum_{j=1}^{n+m} t_j = 1.$$

Dividing each inequality in (1.19) by $t_1 + \cdots + t_n > 0$ and letting

$$\frac{t_1}{t_1 + \cdots + t_n} = y_1, \qquad \frac{t_2}{t_1 + \cdots + t_n} = y_2, \qquad \cdots, \qquad \frac{t_n}{t_1 + \cdots + t_n} = y_n,$$

we obtain

$$\sum_{j=1}^{n} a_{ij} y_j = a_{i1} y_1 + a_{i2} y_2 + \cdots + a_{in} y_n \leq 0, \qquad i = 1, \ldots, m.$$

(2) $0 \notin H$. By Lemma 1, there exists $s = (s_1, \ldots, s_m) \in H$ such that

$$s \cdot a^{(j)} = s_1 a_{1j} + s_2 a_{2j} + \cdots + s_m a_{mj} > 0, \qquad j = 1, \ldots, n, \qquad (1.20)$$
$$s \cdot e^{(i)} = s_i > 0, \qquad i = 1, \ldots, m.$$

Dividing each inequality in (1..20) by $s_1 + \cdots + s_m > 0$ and letting

$$\frac{s_1}{s_1 + \cdots + s_m} = x_1, \qquad \frac{s_2}{s_1 + \cdots + s_m} = x_2, \qquad \cdots, \qquad \frac{s_m}{s_1 + \cdots + s_m} = x_m,$$

we obtain

$$\sum_{i=1}^{m} a_{ij}x_i = a_{1j}x_1 + a_{2j}x_2 + \cdots + a_{mj}x_m > 0, \qquad j = 1, \ldots, n. \qquad \square$$

Theorem 1.3 (The minimax theorem)
Let the payoff matrix of a matrix game be $A = (a_{ij})$. Then

$$v_1 = \max_{X \in S_m} \min_{Y \in S_n} \sum_{i=1}^{m} \sum_{j=1}^{n} a_{ij}x_i y_j = \min_{Y \in S_n} \max_{X \in S_m} \sum_{i=1}^{m} \sum_{j=1}^{n} a_{ij}x_i y_j = v_2. \quad (1.21)$$

Proof. We have proved $v_1 \leq v_2$ in Theorem 1.2, so it is sufficient to give a proof for $v_1 \geq v_2$.

By Lemma 2, one of the following two statements must hold.

(1) There exist $y_1, \ldots, y_n \geq 0$, $\sum_{j=1}^{n} y_j = 1$, such that

$$\sum_{j=1}^{n} a_{ij}y_j \leq 0, \qquad i = 1, \ldots, m.$$

Hence, for any $X = (x_1, \ldots, x_m) \in S_m$,

$$\sum_{i=1}^{m} \left(\sum_{j=1}^{n} a_{ij}y_j \right) x_i \leq 0.$$

Therefore

$$\max_{X \in S_m} \sum_{i=1}^{m} \sum_{j=1}^{n} a_{ij}x_i y_j \leq 0.$$

It follows that

$$v_2 = \min_{Y \in S_n} \max_{X \in S_m} \sum_{i=1}^{m} \sum_{j=1}^{n} a_{ij}x_i y_j \leq 0. \qquad (1.22)$$

(2) There exist $x_1, \ldots, x_m \geq 0$, $\sum_{i=1}^{m} x_i = 1$, such that

$$\sum_{i=1}^{m} a_{ij}x_i > 0, \qquad j = 1, \ldots, n.$$

Hence, for any $Y = (y_1, \ldots, y_n) \in S_n$,

$$\sum_{j=1}^{n} \left(\sum_{i=1}^{m} a_{ij}x_i \right) y_j \geq 0.$$

Therefore,

$$\min_{Y \in S_n} \sum_{i=1}^{m} \sum_{j=1}^{n} a_{ij}x_i y_j \geq 0.$$

It follows that

$$v_1 = \max_{X \in S_m} \min_{Y \in S_n} \sum_{i=1}^{m} \sum_{j=1}^{n} a_{ij} x_i y_j \geqslant 0. \qquad (1.23)$$

By (1.22) and (1.23),

$$\text{either} \quad v_1 \geqslant 0 \qquad \text{or} \quad v_2 \leqslant 0,$$

i.e.

$$\text{never} \quad v_1 < 0 < v_2.$$

Now replace the matrix $A = (a_{ij})$ by

$$(a_{ij} - k) = \begin{bmatrix} a_{11} - k & a_{12} - k & \cdots & a_{1n} - k \\ a_{21} - k & a_{22} - k & \cdots & a_{2n} - k \\ \cdots\cdots\cdots\cdots\cdots\cdots\cdots\cdots\cdots\cdots\cdots \\ a_{m1} - k & a_{m2} - k & \cdots & a_{mn} - k \end{bmatrix},$$

where k is an arbitrary number, and repeat the above arguments. We have

$$\text{never} \quad v_1 - k < 0 < v_2 - k,$$

or

$$\text{never} \quad v_1 < k < v_2. \qquad (1.24)$$

Therefore, $v_1 < v_2$ is impossible, for otherwise there would be a number k satisfying $v_1 < k < v_2$, thus contradicting (1.24).

We have proved $v_1 \geqslant v_2$. $\qquad\qquad\qquad\qquad\qquad\qquad\qquad\qquad$ □

1.6   Inductive proof of the minimax theorem

In this section, we give another proof—an inductive proof—of the minimax theorem; cf. Wang (1982, 1983), Loomis (1946).

We first derive a relation between the minimum (maximum) over mixed strategies and the minimum (maximum) over pure strategies for a set of numbers.

Let $c_1, \ldots, c_n$ be a set of numbers and let $Y = (y_1, \ldots, y_n)$ be any mixed strategy. Suppose

$$\min_{1 \leqslant j \leqslant n} c_j = c_l.$$

Then

$$c_j \geqslant c_l, \qquad j = 1, \ldots, n;$$

hence

$$c_j y_j \geqslant c_l y_j, \qquad j = 1, \ldots, n.$$

For every $Y \in S_n$ we have

$$\sum_{j=1}^{n} c_j y_j \geqslant \sum_{j=1}^{n} c_l y_j = c_l.$$

Hence

$$\min_{Y \in S_n} \sum_{j=1}^{n} c_j y_j \geqslant c_l. \tag{1.25}$$

On the other hand, $Y = (0, 0, \ldots, 0, 1, 0, \ldots, 0)$, where the lth component is 1, is a special mixed strategy. We have

$$c_l = c_1 \cdot 0 + c_2 \cdot 0 + \cdots + c_l \cdot 1 + \cdots + c_n \cdot 0 \geqslant \min_{Y \in S_n} \sum_{j=1}^{n} c_y y_y. \tag{1.26}$$

It follows from (1.25) and (1.26) that

$$\min_{Y \in S_n} \sum_{j=1}^{n} c_j y_j = c_l = \min_{1 \leqslant j \leqslant n} c_j. \tag{1.27}$$

Similarly,

$$\max_{X \in S_m} \sum_{i=1}^{m} d_i x_i = \max_{1 \leqslant i \leqslant m} d_i. \tag{1.28}$$

Using (1.27) and (1.28) we can rewrite v_1 and v_2 as follows:

$$v_1 = \max_{X \in S_m} \min_{Y \in S_n} \sum_{j=1}^{n} \left(\sum_{i=1}^{m} a_{ij} x_i \right) y_j = \max_{X \in S_m} \min_{1 \leqslant j \leqslant n} \sum_{i=1}^{m} a_{ij} x_i; \tag{1.29}$$

$$v_2 = \min_{Y \in S_n} \max_{X \in S_m} \sum_{i=1}^{m} \left(\sum_{j=1}^{n} a_{ij} y_j \right) x_i = \min_{Y \in S_n} \max_{1 \leqslant i \leqslant m} \sum_{j=1}^{n} a_{ij} y_j. \tag{1.30}$$

Thus the minimax theorem can be stated as follows.

Theorem 1.4 (The minimax theorem)

$$\max_{X \in S_m} \min_{1 \leqslant j \leqslant n} \sum_{i=1}^{m} a_{ij} x_i = \min_{Y \in S_n} \max_{1 \leqslant i \leqslant m} \sum_{j=1}^{n} a_{ij} y_j, \tag{1.31}$$

where (a_{ij}) is an arbitrary $m \times n$ matrix. S_m and S_n are respectively sets of points $X = (x_1, \ldots, x_m)$ and $Y = (y_1, \ldots, y_n)$ satisfying

$$x_i \geqslant 0, \qquad i = 1, \ldots, m, \qquad \sum_{i=1}^{m} x_i = 1,$$

$$y_j \geqslant 0, \qquad j = 1, \ldots, n, \qquad \sum_{j=1}^{n} y_j = 1.$$

Proof. By induction. It is obvious that the theorem is true for $m = n = 1$. Assume that the theorem holds for all $(m' < m, n)$, let us prove that it is true for (m, n). (In a similar manner it can be shown that if the theorem holds for all $(m, n' < n)$, then it is true for (m, n).)

Denote

$$v_1^{(m,n)} = \max_{X \in S_m} \min_{1 \leq j \leq n} \sum_{i=1}^{m} a_{ij} x_i, \tag{1.32}$$

$$v_2^{(m,n)} = \min_{Y \in S_n} \max_{1 \leq i \leq m} \sum_{j=1}^{n} a_{ij} y_j. \tag{1.33}$$

Let

$$v_2^{(m,n)} = \min_{Y \in S_n} \max_{1 \leq i \leq m} \sum_{j=1}^{n} a_{ij} y_j = \max_{1 \leq i \leq m} \sum_{j=1}^{n} a_{ij} y_j^*, \tag{1.34}$$

where $(y_1^*, \ldots, y_n^*) = Y^* \in S_n$. Then

$$\sum_{j=1}^{n} a_{ij} y_j^* \leq v_2^{(m,n)}, \qquad i = 1, \ldots, m. \tag{1.35}$$

If the equality holds in (1.35) for all $i = 1, \ldots, m$, as well as in an analogous greater-than-or-equal-to formula corresponding to (1.32) for all $j = 1, \ldots, n$, the validity of the theorem is easily proved. Thus we may assume without loss of generality that

$$\sum_{j=1}^{n} a_{ij} y_j^* = v_2^{(m,n)}, \qquad i = 1, \ldots, m', \tag{1.36}$$

$$\sum_{j=1}^{n} a_{ij} y_j^* < v_2^{(m,n)}, \qquad j = m' + 1, \ldots, m, \tag{1.37}$$

where $m' < m$.

Now consider the reduced matrix game $(i = 1, \ldots, m'; j = 1, \ldots, n)$. We propose to prove the following sequence of inequalities:

$$v_2^{(m',n)} = v_1^{(m',n)} \leq v_1^{(m,n)} \leq v_2^{(m,n)} \leq v_2^{(m',n)}. \tag{1.38}$$
$$\text{(a)} \qquad \text{(b)} \qquad \text{(c)} \qquad \text{(d)}$$

Here (a) is the inductive hypothesis; (c) is evidently valid. We need to prove the inequalities (b) and (d).

To prove (b), let

$$S_{m'} = \{X = (x_1, \ldots, x_{m'})\},$$
$$S_{m'}^0 = \{X = (x_1, \ldots, x_{m'}, 0, \ldots, 0)\} \subset S_m,$$

where $x_i \geq 0$, $\sum x_i = 1$. Then

$$v_1^{(m',n)} = \max_{X \in S_{m'}} \min_{1 \leq j \leq n} \sum_{i=1}^{m'} a_{ij}x_i$$

$$= \min_{1 \leq j \leq n} \sum_{i=1}^{m'} a_{ij}x_i^* \qquad [X' = (x_1^*, \ldots, x_{m'}^*) \in S_{m'}]$$

$$= \min_{1 \leq j \leq n} \sum_{i=1}^{m} a_{ij}x_i^* \qquad [X^* = (x_1^*, \ldots, x_{m'}^*, 0, \ldots, 0) \in S_{m'}^0 \subset S_m]$$

$$\leq \max_{X \in S_m} \min_{1 \leq j \leq n} \sum_{i=1}^{m} a_{ij}x_i = v_1^{(m,n)}.$$

The inequality (b) is proved.

We now proceed to prove the inequality (d), i.e.

$$v_2^{(m,n)} \leq v_2^{(m',n)}.$$

Suppose for the reduced matrix game that

$$v_2^{(m',n)} = \min_{Y \in S_n} \max_{1 \leq i \leq m'} \sum_{j=1}^{n} a_{ij}y_j = \max_{1 \leq i \leq m'} \sum_{j=1}^{n} a_{ij}y_j', \qquad (1.39)$$

where $(y_1', \ldots, y_n') = Y' \in S_n$. Let

$$Y'' = \alpha Y' + (1 - \alpha)Y^* \in S_n, \qquad 0 < \alpha < 1, \qquad (1.40)$$

where $Y'' = (y_1'', \ldots, y_n'')$. Then

$$\sum_{j=1}^{n} a_{ij}y_j'' = \alpha \sum_{j=1}^{n} a_{ij}y_j' + (1 - \alpha) \sum_{j=1}^{n} a_{ij}y_j^*, \qquad i = 1, \ldots, m'. \quad (1.41)$$

Taking $\max_{1 \leq i \leq m'}$ on both sides of (1.41) and utilizing (1.39) and (1.36), we obtain

$$\max_{1 \leq i \leq m'} \sum_{j=1}^{n} a_{ij}y_j'' \leq \alpha v_2^{(m',n)} + (1 - \alpha)v_2^{(m,n)}. \qquad (1.42)$$

(In reality the equality holds in (1.42), because the last term on the right-hand side of (1.41) is a constant.)

Now, for the Y'' in (1.40), it follows from (1.37) and the continuity of the function involved that

$$\sum_{j=1}^{n} a_{ij}y_j'' < v_2^{(m,n)}, \qquad i = m' + 1, \ldots, m \qquad (1.43)$$

if α is sufficiently small. But

$$\max_{1\leq i\leq m} \sum_{j=1}^{n} a_{ij}y_j'' \geq \min_{Y\in S_n} \max_{1\leq i\leq m} \sum_{j=1}^{n} a_{ij}y_j = v_2^{(m,n)} \tag{1.44}$$

by (1.33). It follows from (1.43) and (1.44) that

$$\max_{1\leq i\leq m'} \sum_{j=1}^{n} a_{ij}y_j'' \geq v_2^{(m,n)}.$$

Thus we may replace the left-hand side of (1.42) by $v_2^{(m,n)}$ to obtain

$$\alpha v_2^{(m',n)} \geq v_2^{(m,n)} - (1-\alpha)v_2^{(m,n)} = \alpha v_2^{(m,n)},$$

or

$$v_2^{(m,n)} \leq v_2^{(m',n)}.$$

We have proved the validity of the inequalities (1.38), therefore,

$$v_1^{(m,n)} = v_2^{(m,n)}. \qquad\qquad \square$$

1.7   Saddle points in mixed strategies

Let $A = (a_{ij})$ be the payoff matrix of an $m \times n$ matrix game. If $X = (x_1, \ldots, x_m) \in S_m$ and $Y = (y_1, \ldots, y_n) \in S_n$ are respectively mixed strategies of players 1 and 2, then the expected payoff

$$\sum_{i=1}^{m} \sum_{j=1}^{n} a_{ij}x_i y_j$$

can be written in matrix notation

$$\sum_{i=1}^{m} \sum_{j=1}^{n} a_{ij}x_i y_j = XAY^t,$$

where Y^t is the transpose of Y.

Definition 1.2. Suppose $X^* \in S_m$, $Y^* \in S_n$. (X^*, Y^*) is called a *saddle point* (in mixed strategies) of the matrix game $A = (a_{ij})$ if

$$XAY^{*t} \leq X^*AY^{*t} \leq X^*AY^t \tag{1.45}$$

for all $X \in S_m$ and all $Y \in S_n$.

The following theorem establishes the equivalence of the existence of a saddle point and the minimax theorem.

Theorem 1.5. *A necessary and sufficient condition for the $m \times n$ matrix* game $A^* = (a_{ij})$ *to have a saddle point is that*

$$\max_{X \in S_m} \min_{Y \in S_n} XAY^t \quad \text{and} \quad \min_{Y \in S_n} \max_{X \in S_m} XAY^t \quad (1.46)$$

exist and be equal.

Proof. *Necessity.* It is obvious that the two quantities in (1.46) both exist. Assume that XAY^t has a saddle point (X^*, Y^*). That is to say, the inequalities

$$XAY^{*t} \leqslant X^*AY^{*t} \leqslant X^*AY^t \quad (1.47)$$

hold for all $X \in S_m$ and all $Y \in S_n$. From the first inequality in (1.47), we have

$$\max_{X \in S_m} XAY^{*t} \leqslant X^*AY^{*t};$$

hence

$$\min_{Y \in S_n} \max_{X \in S_m} XAY^t \leqslant X^*AY^{*t}. \quad (1.48)$$

Similarly, from the second inequality in (1.47), we have

$$X^*AY^{*t} \leqslant \min_{Y \in S_n} X^*AY^t \leqslant \max_{X \in S_m} \min_{Y \in S_n} XAY^t. \quad (1.49)$$

It follows from (1.48) and (1.49) that

$$\min_{Y \in S_n} \max_{X \in S_m} XAY^t \leqslant \max_{X \in S_m} \min_{Y \in S_n} XAY^t.$$

But it is known that the reverse inequality holds (see Theorem 1.2). Therefore,

$$\max_{X \in S_m} \min_{Y \in S_n} XAY^t = \min_{Y \in S_n} \max_{X \in S_m} XAY^t,$$

and the necessity of the condition is proved.

Sufficiency. Assume that the two values in (1.46) are equal. Let

$$\max_{X \in S_m} \min_{Y \in S_n} XAY^t = \min_{Y \in S_n} X^*AY^t, \quad (1.50)$$

$$\min_{Y \in S_n} \max_{X \in S_m} XAY^t = \max_{X \in S_m} XAY^{*t}. \quad (1.51)$$

By the definitions of minimum and maximum,

$$\min_{Y \in S_n} X^*AY^t \leqslant X^*AY^{*t}, \quad (1.52)$$

$$X^*AY^{*t} \leqslant \max_{X \in S_m} XAY^{*t}. \quad (1.53)$$

Since the left-hand sides of (1.50) and (1.51) are equal, all terms in (1.50) through (1.53) are equal to each other. In particular, we have

$$\max_{X \in S_m} XAY^{*t} = X^*AY^{*t}.$$

Therefore,

$$XAY^{*t} \leqslant X^*AY^{*t} \tag{1.54}$$

for all $X \in S_m$. Similarly,

$$X^*AY^{*t} \leqslant X^*AY^t \tag{1.55}$$

for all $Y \in S_n$. By (1.54) and (1.55), (X^*, Y^*) is a saddle point of XAY^t.

$$\square$$

1.8   Optimal strategies and their properties

If the matrix game $A = (a_{ij})$ has a saddle point (it has been shown in the preceding section that a saddle point always exists) and if (X^*, Y^*) is a saddle point, i.e.

$$XAY^{*t} \leqslant X^*AY^{*t} \leqslant X^*AY^t \tag{1.45}$$

for all $X \in S_m$ and all $Y \in S_n$, then we say that X^*, Y^* are respectively *optimal strategies* of players 1 and 2, and

$$X^*AY^{*t}$$

is the *value* of the game. That is to say, the expected payoff of the game at the saddle point (X^*, Y^*) is the value of the game. We also say that (X^*, Y^*) is a *solution* of the game. By Theorem 1.5 the value of the game is the common value of

$$v_1 = \max_{X \in S_m} \min_{Y \in S_n} XAY^t \quad \text{and} \quad v_2 = \min_{Y \in S_n} \max_{X \in S_m} XAY^t.$$

The definition (1.45) of a saddle point shows that, as long as player 1 sticks to his optimal strategy X^*, he can be sure to get at least the expected payoff X^*AY^{*t} no matter which strategy player 2 chooses; similarly, as long as player 2 sticks to his optimal strategy Y^*, he can hold player 1's expected payoff down to at most X^*AY^{*t} no matter how player 1 makes his choice of strategy.

The following notations will be used. We denote the ith row vector of the matrix A by $A_i.$ and the jth column vector of A by $A_{.j}$. Thus

$$XA_{.j} = \sum_{i=1}^{m} a_{ij} x_i, \qquad A_i. Y^t = \sum_{j=1}^{n} a_{ij} y_j.$$

$XA_{.j}$ is the expected payoff when player 1 chooses the mixed strategy X and player 2 chooses the pure strategy j; $A_i.Y^t$ is the expected payoff when player 2 chooses the mixed strategy Y and player 1 the pure strategy i.

We now give some essential propeties of optimal strategies.

Theorem 1.6. *Let $A = (a_{ij})$ be the payoff matrix of an $m \times n$ matrix game whose value is v.*

(1) *Let Y^* be an optimal strategy of player 2. If*

$$A_i.Y^{*t} < v,$$

then $x_i^* = 0$ in every optimal strategy X^* of player 1.

(2) *Let X^* be an optimal strategy of player 1. If*

$$X^*A_{.j} > v,$$

then $y_j^* = 0$ in every optimal strategy Y^* of player 2.

Proof. Only (1) will be proved. The proof of (2) is similar. Since Y^* is an optimal strategy of player 2, we have

$$A_i.Y^{*t} \leq v, \qquad i = 1, \ldots, m.$$

Let

$$S_1 = \{i: A_i.Y^{*t} < v\}, \qquad S_2 = \{i: A_i.Y^{*t} = v\}.$$

Then

$$v = X^*AY^{*t} = \sum_{i=1}^{m} x_i^* A_i.Y^{*t}$$

$$= \sum_{i \in S_1} x_i^* A_i.Y^{*t} + \sum_{i \in S_2} x_i^* A_i.Y^{*t}$$

$$= \sum_{i \in S_1} x_i^* A_i.Y^{*t} + \sum_{i \in S_2} x_i^* v.$$

Hence

$$v\left(1 - \sum_{i \in S_2} x_i^*\right) = \sum_{i \in S_1} x_i^* A_i.Y^{*t},$$

i.e.

$$v\sum_{i \in S_1} x_i^* = \sum_{i \in S_1} x_i^* A_i.Y^{*t} \quad \text{or} \quad \sum_{i \in S_1} (v - A_i.Y^{*t})x_i^* = 0.$$

Since $i \in S_1$ implies $v - A_i.Y^{*t} > 0$, we must have $x_i^* = 0$. $\qquad \square$

This theorem states that if player 2 has an optimal strategy Y^* in a matrix game with value v, and if player 1 cannot by using the ith pure strategy attain the expected payoff v, then the pure strategy i is a bad strategy and cannot appear in any of his optimal mixed strategies.

Theorem 1.7. *Let $A = (a_{ij})$ be the payoff matrix of an $m \times n$ matrix game whose value is v.*

  (1) *A necessary and sufficient condition for $X^* \in S_m$ to be an optimal strategy of player 1 is that*

$$v \leqslant X^* A_{\cdot j}, \qquad j = 1, \ldots, n.$$

  (2) *A necessary and sufficient condition for $Y^* \in S_n$ to be an optimal strategy of player 2 is that*

$$A_i . Y^{*t} \leqslant v, \qquad i = 1, \ldots, m.$$

Proof. Only (1) will be proved. The proof of (2) is similar.. Necessity of the condition follows directly from the definition of a saddle point.

To prove the sufficiency of the condition, assume that

$$v \leqslant X^* A_{\cdot j}, \qquad j = 1, \ldots, n. \tag{1.56}$$

Let (X^0, Y^0) be a saddle point of the game, i.e.

$$XAY^{0t} \leqslant X^0 A Y^{0t} \leqslant X^0 A Y^t \tag{1.57}$$

for all $X \in S_m$ and all $Y \in S_n$.

We will prove that (X^*, Y^0) is a saddle point of the game. Let $Y = (y_1, \ldots, y_n) \in S_n$ be any mixed strategy of player 2. Multiplying both sides of (1.56) by y_j and summing for $j = 1, \ldots, n$ we obtain

$$v \leqslant \sum_{j=1}^{n} X^* A_{\cdot j} y_j = X^* A Y^t. \tag{1.58}$$

In particular,

$$v \leqslant X^* A Y^{0t}. \tag{1.59}$$

The definition of saddle point (1.57) implies

$$X^* A Y^{0t} \leqslant X^0 A Y^{0t} = v. \tag{1.60}$$

From (1.59) and (1.60) we have

$$X^* A Y^{0t} = X^0 A Y^{0t} = v. \tag{1.61}$$

It follows from (1.57), (1.61) and (1.58) that

$$XAY^{0t} \leqslant X^* A Y^{0t} \leqslant X^* A Y^t,$$

which proves that (X^*, Y^0) is a saddle point of the game. Hence, X^* is an optimal strategy of player 1. $\qquad \square$

If the value of a game is known, the above theorem can be used to examine whether a given strategy $X^*(Y^*)$ of player 1(2) is optimal.

1.9 Domination of strategies

Consider the matrix game whose payoff matrix is

$$\begin{bmatrix} 1 & -2 & 0 \\ 0 & 1 & -1 \\ 2 & -1 & 1 \end{bmatrix}. \tag{1.62}$$

An examination of the elements of the payoff matrix shows that player 1 will never use his first strategy since each element in the first row is smaller than the corresponding element in the third row. Hence, regardless of which strategy player 2 chooses, player 1 will gain more by choosing strategy 3 than by choosing strategy 1. Strategy 1 of player 1 can only appear in his optimal mixed strategies with probability zero.

Thus, in order to solve the matrix game (1.62), the first row can be deleted and we need only consider the resulting matrix

$$\begin{bmatrix} 0 & 1 & -1 \\ 2 & -1 & 1 \end{bmatrix}. \tag{1.63}$$

Now in this matrix each element of the first column is greater than the corresponding element of the third column. So player 2 will lose less by choosing strategy 3 than by choosing strategy 1. Thus the first strategy of player 2 will never be included in any of his optimal mixed strategies with positive probability.

Therefore the first column of the matrix (1.63) can be deleted to obtain

$$\begin{bmatrix} 1 & -1 \\ -1 & 1 \end{bmatrix}. \tag{1.64}$$

It is easily verified that this 2×2 matrix game has the mixed strategy solution: $X^* = (\frac{1}{2}, \frac{1}{2})$, $Y^* = (\frac{1}{2}, \frac{1}{2})$, and $v = 0$.

Returning to the original 3×3 matrix game (1.62), its solution is obviously

$$X^* = (0, \tfrac{1}{2}, \tfrac{1}{2}), \qquad Y^* = (0, \tfrac{1}{2}, \tfrac{1}{2}); \qquad v = 0.$$

We have seen that in (1.62) player 1 will never use his stragegy 1 since strategy 3 gives him a greater payoff than strategy 1. Similarly, in (1.63) player 2 will never use his strategy 1 since it always costs him a greater loss than strategy 3.

We now give the definition of domination of strategies.

Definition 1.3. Let $A = (a_{ij})$ be the payoff matrix of an $m \times n$ matrix game. If

$$a_{kj} \geqslant a_{lj}, \qquad j = 1, \ldots, n, \tag{1.65}$$

we say that player 1's strategy k *dominates* strategy l. If

$$a_{ik} \leqslant a_{il}, \qquad i = 1, \ldots, m, \qquad (1.66)$$

we say that player 2's strategy k dominates strategy l.

If the inequalities in (1.65) or (1.66) are replaced by strict inequalities, we say that the strategy k of player 1 or 2 strictly dominates his strategy l.

The concept of domination for mixed strategies is similar.

Let us consider the case in which a pure strategy is dominated by a convex linear combination of several other pure strategies. It can be proved that in this case if the domination is strict, then we can delete the row or column in the payoff matrix corresponding to the dominated pure strategy and solve the reduced matrix game. The optimal strategies of the original matrix game can be obtained from those of the reduced one by assigning the probability zero to the pure strategy corresponding to the deleted row or column.

If the domination is not strict, we can still obtain a solution for the original game from that of the reduced game by the procedure described in the preceding paragraph. But in this situation the deletion of a row or column may involve loss of some optimal strategies of the original game. That is to say, if non-strict domination of strategies is used to delete a row or a column, a complete set of optimal strategies for a reduced game will not necessarily lead to a complete set of optimal strategies for the original game. Of course, if what we need is any one of the mixed strategy saddle points, and in practice this is usually the case, the above procedure will suffice.

Example 1.5. Let the payoff matrix of a matrix game be

$$\begin{bmatrix} 3 & 2 & 4 & 0 \\ 3 & 4 & 2 & 3 \\ 4 & 3 & 4 & 2 \\ 0 & 4 & 0 & 8 \end{bmatrix}.$$

Strategy 1 of player 1 is dominated by his strategy 3, so we can delete the first row of the payoff matrix:

$$\begin{bmatrix} 3 & 4 & 2 & 3 \\ 4 & 3 & 4 & 2 \\ 0 & 4 & 0 & 8 \end{bmatrix}.$$

Strategy 1 of player 2 is dominated by his strategy 3, so the first column can be deleted:

$$\begin{bmatrix} 4 & 2 & 3 \\ 3 & 4 & 2 \\ 4 & 0 & 8 \end{bmatrix}.$$

It is easily seen that the elements of this 3×3 matrix satisfy

$$\begin{bmatrix} 4 \\ 3 \\ 4 \end{bmatrix} \geqslant \frac{1}{2} \begin{bmatrix} 2 \\ 4 \\ 0 \end{bmatrix} + \frac{1}{2} \begin{bmatrix} 3 \\ 2 \\ 8 \end{bmatrix}.$$

Hence the first column of the 3×3 matrix can be deleted to obtain

$$\begin{bmatrix} 2 & 3 \\ 4 & 2 \\ 0 & 8 \end{bmatrix}.$$

The first row of this 3×2 matrix is dominated by a convex linear combination of the second and third rows:

$$(2 \ \ 3) \leqslant \tfrac{1}{2}(4 \ \ 2) + \tfrac{1}{2}(0 \ \ 8).$$

Hence we can delete the first row of the 3×2 matrix to obtain

$$\begin{bmatrix} 4 & 2 \\ 0 & 8 \end{bmatrix}.$$

It is easy to verify that the optimal strategies of this 2×2 matrix game are $X^* = (\tfrac{4}{5}, \tfrac{1}{5})$, $Y^* = (\tfrac{3}{5}, \tfrac{2}{5})$ and that the value is $\tfrac{16}{5}$.

The 2×2 matrix is the submatrix of the original 4×4 matrix formed by its third, fourth rows and third, fourth columns. Therefore,

$$X^* = (0, 0, \tfrac{4}{5}, \tfrac{1}{5}), \qquad Y^* = (0, 0, \tfrac{3}{5}, \tfrac{2}{5})$$

are optimal strategies of the original matrix game, and the value is $\tfrac{16}{5}$.

Example 1.6. Let the payoff matrix of a matrix game be

$$\begin{bmatrix} 3 & 5 & 3 \\ 4 & -3 & 2 \\ 3 & 2 & 3 \end{bmatrix}.$$

This game has a saddle point at $i = 1, j = 3$. Hence $(1, 3)$ is a solution of the game and the value is $a_{13} = 3$.

If domination of strategies is considered, we can first delete the third row of this matrix and then delete the first column in the remaining 2×3

matrix, and the result is the 2×2 matrix game

$$\begin{bmatrix} 5 & 3 \\ -3 & 2 \end{bmatrix}.$$

Here the upper right element is the row minimum and at the same time the column maximum, so there is still a saddle point $(1, 2)$ in this 2×2 game. For the original 3×3 matrix game, we have again the optimal pure strategies

$$X_1^* = (1, 0, 0), \qquad Y_1^* = (0, 0, 1).$$

However, it turns out that the mixed strategies

$$X_2^* = (\tfrac{1}{3}, 0, \tfrac{2}{3}), \qquad Y_2^* = (\tfrac{1}{2}, 0, \tfrac{1}{2})$$

are also optimal for players 1 and 2, respectively. This can be easily verified using Theorem 1.7. See Example 1.7 (Section 1.12).

In reality, all convex linear combinations of X_1^* and X_2^* are optimal strategies of player 1; all convex linear combinations of Y_1^* and Y_2^* are optimal strategies of player 2.

This example shows that, as was mentioned above, the deletion of certain row or column of a payoff matrix using non-strict domination of strategies may result in a smaller game whose complete set of solutions does not lead to the complete set of solutions of the original larger game. That is, the solution procedure may lose some optimal strategies of the original game.

1.10   Solution of 2×2 matrix games

Let the payoff matrix of a 2×2 matrix game be

$$\begin{bmatrix} a & b \\ c & d \end{bmatrix}. \tag{1.67}$$

If a saddle point exists, we at once get the pure strategy solution. In the case that there does not exist a saddle point, by exchanging rows or columns of the matrix, i.e. by renumbering the two strategies of player 1 or 2, it can be seen that only the following case needs to be considered:

$$a < b, \qquad a < c, \qquad d < b, \qquad d < c.$$

In this case, the game must have mixed strategy solution.

Let $X^* = (x^*, 1 - x^*)$, $Y^* = (y^*, 1 - y^*)$ be respectively optimal strategies of players 1 and 2, where

$$0 < x^* < 1, \qquad 0 < y^* < 1. \tag{1.68}$$

Since
$$x^* > 0, \qquad 1 - x^* > 0, \qquad y^* > 0, \qquad 1 - y^* > 0,$$

by Theorem 1.6 the following equations hold:
$$X^*A_{.1} = v, \qquad X^*A_{.2} = v, \qquad A_{1.}Y^{*t} = v, \qquad A_{2.}Y^{*t} = v,$$

where v is the value of the game. Writing these equations in terms of the elements of the matrix, we have
$$ax^* + c(1 - x^*) = v, \qquad bx^* + d(1 - x^*) = v,$$
$$ay^* + b(1 - y^*) = v, \qquad cy^* + d(1 - y^*) = v.$$

The first two equations give

$$x^* = \frac{d - c}{a + d - b - c}, \tag{1.69}$$

and the last two equations give

$$y^* = \frac{d - b}{a + d - b - c}. \tag{1.70}$$

Thus

$$v = \frac{ad - bc}{a + d - b - c}. \tag{1.71}$$

Equations (1.69), (1.70), (1.71) are the optimal strategies and the value of the game (1.67) when it has no saddle point in pure strategies.

These formulae are also valid for the case
$$a > b, \qquad a > c, \qquad d > b, \qquad d > c.$$

For the 2×2 matrix game with no saddle point, an interesting technique of solution is described in Williams (1954). The method is very easily memorized. For example, let us solve the matrix game

$$A = \begin{bmatrix} 1 & 4 \\ 3 & -2 \end{bmatrix}.$$

First, substract each element of the second column from the corresponding element of the first column:

$$\left.\begin{array}{c} 1 - 4 \\ 3 - (-2) \end{array}\right\} = \left\{\begin{array}{c} -3 \\ 5 \end{array}\right..$$

Then take absolute values of the two differences and reverse the order of the absolute values:

$$\left.\begin{array}{c} -3 \\ 5 \end{array}\right\} \rightarrow \left\{\begin{array}{c} 5 \\ 3 \end{array}\right..$$

The ratio $5:3$ is the ratio of x_1 to x_2 in player 1's optimal strategy

$$X^* = (x_1, x_2) = (x^*, 1 - x^*).$$

We obtain

$$X^* = (\tfrac{5}{8}, \tfrac{3}{8}).$$

Similarly, subtract each element of the second row from the corresponding element of the first row:

Then take absolute values of the two differences and reverse the order of the absolute values:

The ratio $6:2$ is the ratio of y_1 to y_2 in player 2's optimal strategy

$$Y^* = (y_1, y_2) = (y^*, 1 - y^*).$$

We have

$$Y^* = (\tfrac{3}{4}, \tfrac{1}{4}).$$

The value of the game is easily calculated:

$$v = X^* A Y^{*t} = \tfrac{7}{4}.$$

The results are the same if we evaluate X^*, Y^* and v by the formulae (1.69), (1.70), (1.71).

1.11 Graphical solution of $2 \times n$ and $m \times 2$ matrix games

We will illustrate the method by a 2×3 matrix game. Suppose the payoff matrix A is

		$\boxed{1}$	$\boxed{2}$	$\boxed{3}$
x	①	a	b	c
$1-x$	②	d	e	f

Denote player 1's pure strategies by ①, ② and player 2's pure strategies by $\boxed{1}$, $\boxed{2}$, $\boxed{3}$.

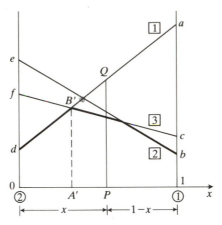

Figure 1.1

Assume that player 1 uses the mixed strategy

$$X = (x_1, x_2) = (x, 1 - x),$$

where $0 \leqslant x \leqslant 1$. $x = 1$ represents the pure strategy ①, $x = 0$ represents the pure strategy ②.

If player 1 chooses the pure strategy ①, i.e. when $x = 1$, and if player 2 chooses the pure strategy ①, the payoff is a, as shown in Fig. 1.1. If player 1 chooses ②, i.e. when $x = 0$, the payoff corresponding to ① is d. We join the line ad in the figure.

Now suppose that player 1 chooses a mixed strategy $X = (x, 1 - x)$ represented by P in the figure. Then it can be seen that the height PQ represents the expected payoff when player 1 uses X and player 2 uses ①. The amount is

$$XA_{\cdot 1} = \sum_{i=1}^{2} a_{i1}x_i = ax + d(1 - x).$$

Similarly, corresponding to player 2's strategies ② and ③ we have the lines be and cf. The heights of the points on these lines represent the expected payoffs if player 1 uses X while player 2 uses ② and ③, respectively.

For any mixed strategy X of player 1, his expected payoff is at least the minimum of the three ordinates on the lines ad, be, cf at the point x, i.e.

$$\min_{1 \leqslant j \leqslant 3} XA_{\cdot j} = \min_{1 \leqslant j \leqslant 3} \sum_{i=1}^{2} a_{ij}x_i. \tag{1.72}$$

This function is represented by the heavy black line in the figure.

Player 1 wishes to choose an X so as to maximize the minimum function in (1.72). We see from the figure that he should choose the mixed strategy corresponding to the point A'. At this point the expected payoff is

$$A'B' = \max_{X \in S_2} \min_{1 \leqslant j \leqslant 3} \sum_{i=1}^{2} a_{ij} x_i,$$

which is the value of the game.

Note that the point B' in Fig. 1.1 is the intersection of the lines ad and cf. The abscissa $x = x^*$ of the point A' and the value of $A'B'$ can be evaluated by solving a set of two linear equations in two unknowns.

The graph also shows that player 2's optimal strategy does not involve his pure strategy $\boxed{2}$. Therefore, the solution of the 2×3 matrix game can be obtained from the solution of the 2×2 matrix game

$$\begin{bmatrix} a & c \\ d & f \end{bmatrix}.$$

The graphical method described above can be used to solve all $2 \times n$ matrix games.

In the particular cases, it may happen that the graphical solution obtained for player 1 is a subinterval of the interval $[0, 1]$ on the x-axis, or an end point of $[0, 1]$. The latter case corresponds to a pure strategy solution of player 1. The former is the case in which the maximum of minimum is represented by a horizontal segment contained in the heavy black line in Fig. 1.1.

The graphical solution of an $m \times 2$ matrix game is similar. We now explain it for the case $m = 3$ and let the payoff matrix A of the game be

$$
\begin{array}{cc}
 & y \quad\ 1-y \\
 & \boxed{1} \quad \boxed{2} \\
\begin{array}{c} ① \\ ② \\ ③ \end{array} &
\begin{bmatrix} a & b \\ c & d \\ e & f \end{bmatrix}.
\end{array}
$$

Assume that player 2 uses the mixed strategy

$$Y = (y_1, y_2) = (y, 1-y),$$

where $0 \leqslant y \leqslant 1$. $y = 1$ and $y = 0$ represent the pure strategies $\boxed{1}$ and $\boxed{2}$ respectively.

The ordinate of a point on the heavy black line in Fig. 1.2 is

$$\max_{1 \leqslant i \leqslant 3} A_i . Y^t = \max_{1 \leqslant i \leqslant 3} \sum_{j=1}^{2} a_{ij} y_j.$$

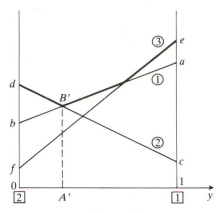

Figure 1.2

Player 2 wishes to choose a Y so as to minimize the above maximum. This Y is represented by the point A'. The expected payoff at this point,

$$A'B' = \min_{Y \in S_2} \max_{1 \leqslant i \leqslant 3} A_i . Y^t,$$

is the value of the game.

1.12 Solution of 3×3 matrix games

We shall use the barycentric coordinates for the points $X = (x_1, x_2, x_3)$ satisfying

$$x_1 \geqslant 0, \qquad x_2 \geqslant 0, \qquad x_3 \geqslant 0, \tag{1.73}$$

$$x_1 + x_2 + x_3 = 1. \tag{1.74}$$

Let 123 be an equilateral triangle with unit perpendicular bisectors. For every point X in this closed triangle, let x_1, x_2, x_3 be the distances from X to the sides of the triangle opposite the vertices 1, 2, 3, respectively. Then x_1, x_2, x_3 satisfy the conditions (1.73) and (1.74); see Fig. 1.3. They are called the *barycentric coordinates* of X. The set of all points in the closed triangle is the simplex S_3.

The barycentric coordinates of the vertices 1, 2 and 3 are $(1, 0, 0)$, $(0, 1, 0)$ and $(0, 0, 1)$, respectively. The equations of the three sides 23, 31, 12 of the triangle are

$$x_1 = 0, \qquad x_2 = 0, \qquad x_3 = 0,$$

respectively.

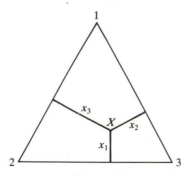

Figure 1.3

Consider the payoff matrix of an arbitrary 3×3 matrix game:

$$A = \begin{bmatrix} a_{11} & a_{12} & a_{13} \\ a_{21} & a_{22} & a_{23} \\ a_{31} & a_{32} & a_{33} \end{bmatrix}.$$

The value of the game is

$$v = \max_{X \in S_3} \min_{1 \leq j \leq 3} XA_{\cdot j} = \max_{X \in S_3} \min \begin{Bmatrix} XA_{\cdot 1} \\ XA_{\cdot 2} \\ XA_{\cdot 3} \end{Bmatrix}. \qquad (1.75)$$

Consider the equations

$$XA_{\cdot 1} = XA_{\cdot 2}, \qquad (1.76)$$
$$XA_{\cdot 2} = XA_{\cdot 3}, \qquad (1.77)$$
$$XA_{\cdot 3} = XA_{\cdot 1}, \qquad (1.78)$$

Each equation represents a straight line which divides the whole plane into two half-planes. (The points outside the triangle can be regarded as points satisfying the condition (1.74) but with one or two of the three coordinates x_1, x_2, x_3 assuming negative values.) Equation (1.76), for instance, divides the whole plane into two half-planes. The points X in one half-plane satisfy the condition

$$XA_{\cdot 1} < XA_{\cdot 2},$$

while those in the other half-plane satisfy the condition

$$XA_{\cdot 1} > XA_{\cdot 2}.$$

The situation is similar for equations (1.77) and (1.78).

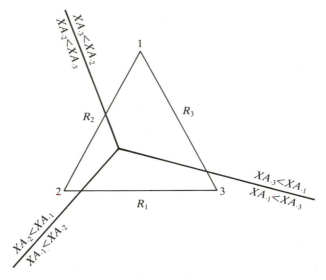

Figure 1.4

The three lines (1.76), (1.77), (1.78) either intersect at one point or are parallel to each other. In both cases these lines divide the whole plane into three regions R_1, R_2, R_3; see Fig. 1.4. In the region R_1,

$$\min_{1\leqslant j\leqslant 3} XA_{.j} = XA_{.1};$$

in R_2,

$$\min_{1\leqslant j\leqslant 3} XA_{.j} = XA_{.2};$$

in R_3,

$$\min_{1\leqslant j\leqslant 3} XA_{.j} = XA_{.3}.$$

Therefore, (1.75) can be written as

$$v = \max_{X\in S_3} \min_{1\leqslant j\leqslant 3} XA_{.j}$$

$$= \max\left\{ \max_{X\in S_3\cap R_1} XA_{.1},\ \max_{X\in S_3\cap R_2} XA_{.2},\ \max_{X\in S_3\cap R_3} XA_{.3}\right\}. \quad (1.79)$$

To determine the value v, we should first compute

$$\max_{X\in S_3\cap R_j} XA_{.j}, \quad j = 1, 2, 3. \quad (1.80)$$

Now each of the sets $S_3 \cap R_j$, $j = 1, 2, 3$, is a convex polygon (in the special case it may be a line segment, a point, or the empty set). A linear function $XA_{.j}$ on a convex polygon can assume its maximum only at a vertex of the polygon. Hence it is sufficient to evaluate the values of $XA_{.j}$ at the relevant vertices and make a comparison between these values. The maximum value must be v. In the course of the comparison, the optimal strategies of player 1 can be determined.

After the value v of the game is determined, the optimal strategies of player 2 can be found out in a similar manner. We have

$$v = \min_{Y \in S_3} \max_{1 \leqslant i \leqslant 3} A_i . Y^t = \min_{Y \in S_3} \max \left\{ \begin{array}{c} A_1 . Y^t \\ A_2 . Y^t \\ A_3 . Y^t \end{array} \right\}$$

$$= \min \left\{ \min_{Y \in S_3 \cap T_1} A_1 . Y^t, \quad \min_{Y \in S_3 \cap T_2} A_2 . Y^t, \quad \min_{Y \in S_3 \cap T_3} A_3 . Y^t \right\},$$

where T_i is the region in which the linear function $A_i . Y^t$ satisfies

$$A_i . Y^t = \max_{1 \leqslant i \leqslant 3} A_i . Y^t$$

for $i = 1, 2, 3$. It suffices to compute the values of $A_i . Y^t$ at the relevant vertices and make a comparison between them. The minimum value must be v, and the vertices Y at which the minimum is assumed are points corresponding to the optimal strategies of player 2.

Example 1.7. Let us compute the value and optimal strategies of the game of Example 1.6. The payoff matrix is

$$B = \begin{bmatrix} 3 & 5 & 3 \\ 4 & -3 & 2 \\ 3 & 2 & 3 \end{bmatrix}.$$

To simplify the computation, we add a constant -3 to each element of the matrix B. The result is

$$A = \begin{bmatrix} 0 & 2 & 0 \\ 1 & -6 & -1 \\ 0 & -1 & 0 \end{bmatrix}.$$

For this matrix game A we have

$$XA_{.1} = x_2, \qquad XA_{.2} = 2x_1 - 6x_2 - x_3, \qquad XA_{.3} = -x_2.$$

The equation of the line $XA_{.1} = XA_{.2}$ is

$$2x_1 - 7x_2 - x_3 = 0, \qquad \text{or} \quad 3x_1 - 6x_2 = 1;$$

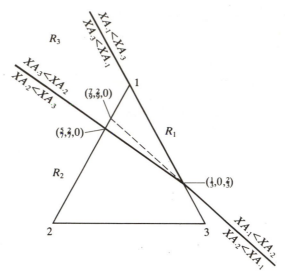

Figure 1.5

that of $XA_{.2} = XA_{.3}$ is

$$2x_1 - 5x_2 - x_3 = 0, \qquad \text{or} \quad 3x_1 - 4x_2 = 1;$$

and that of $XA_{.3} = XA_{.1}$ is

$$x_2 = 0.$$

The regions R_1, R_2, R_3 in which $\min_{1 \le j \le 3} XA_{.j}$ equals $XA_{.1}$, $XA_{.2}$, $XA_{.3}$ respectively are shown in Fig. 1.5.

We evaluate $XA_{.1}$ at the points $(1, 0, 0)$ and $(\frac{1}{3}, 0, \frac{2}{3})$:

$$(1, 0, 0)(0, 1, 0)^t = 0,$$
$$(\tfrac{1}{3}, 0, \tfrac{2}{3})(0, 1, 0)^t = 0.$$

The values of $XA_{.2}$ at $(\frac{5}{7}, \frac{2}{7}, 0)$, $(0, 1, 0)$, $(0, 0, 1)$ are also evaluated:

$$(\tfrac{5}{7}, \tfrac{2}{7}, 0)(2, -6, -1)^t = -\tfrac{2}{7},$$
$$(0, 1, 0)(2, -6, -1)^t = -6,$$
$$(0, 0, 1)(2, -6, -1)^t = -1.$$

Comparing the five equations above, we see that the value of the matrix game A is

$$v_A = 0, \tag{1.81}$$

and the vertices at which the maximum $v_A = 0$ is assumed are

$$X_1^* = (1, 0, 0), \qquad X_2^* = (\tfrac{1}{3}, 0, \tfrac{2}{3}). \qquad (1.82)$$

X_1^* and X_2^* are optimal strategies of player 1.
   Now let us compute optimal strategies of player 2. We have

$$A_1.Y^t = 2y_2, \qquad A_2.Y^t = y_1 - 6y_2 - y_3, \qquad A_3.Y^t = -y_2.$$

The equation of the line $A_1.Y^t = A_2.Y^t$ is

$$y_1 - 8y_2 - y_3 = 0, \qquad \text{or} \quad 2y_1 - 7y_2 = 1;$$

that of $A_2.Y^t = A_3.Y^t$ is

$$y_1 - 5y_2 - y_3 = 0 \qquad \text{or} \quad 2y_1 - 4y_2 = 1;$$

and that of $A_3.Y^t = A_1.Y^t$ is

$$y_2 = 0.$$

   The regions T_1, T_2, T_3 in which $\max\limits_{1 \leqslant i \leqslant 3} A_i.Y^t$ equals $A_1.Y^t$, $A_2.Y^t$, $A_3.Y^t$
respectively are shown in Fig. 1.6.
   The values of $A_1.Y^t$ are the vertices $(0, 1, 0)$, $(0, 0, 1)$, $(\tfrac{1}{2}, 0, \tfrac{1}{2})$, $(\tfrac{8}{9}, \tfrac{1}{9}, 0)$

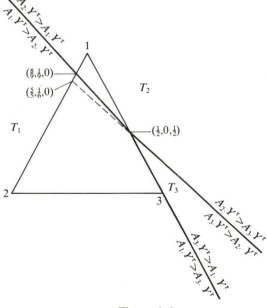

Figure 1.6

are

$$(0, 2, 0) (0, 1, 0)^t = 2,$$
$$(0, 2, 0) (0, 0, 1)^t = 0,$$
$$(0, 2, 0) (\tfrac{1}{2}, 0, \tfrac{1}{2})^t = 0,$$
$$(0, 2, 0) (\tfrac{8}{9}, \tfrac{1}{9}, 0)^t = \tfrac{2}{9}.$$

The value of $A_2.Y^t$ at the vertex $(1, 0, 0)$ is

$$(1, -6, -1) (1, 0, 0)^t = 1.$$

Comparing the five equations above, we see that the vertices

$$Y_1^* = (0, 0, 1), \qquad Y_2^* = (\tfrac{1}{2}, 0, \tfrac{1}{2}) \qquad (1.83)$$

represent optimal strategies of player 2.

Returning to the original matrix game B, we obtain the value of B:

$$v_B = 3.$$

The complete set of optimal strategies for player 1 is

$$\lambda X_1^* + (1 - \lambda)X_2^*, \qquad 0 \leq \lambda \leq 1,$$

i.e.

$$\lambda(1, 0, 0) + (1 - \lambda)(\tfrac{1}{3}, 0, \tfrac{2}{3}), \qquad 0 \leq \lambda \leq 1;$$

the complete set of optimal strategies for player 2 is

$$\mu Y_1^* + (1 - \mu)Y_2^*, \qquad 0 \leq \mu \leq 1,$$

i.e.

$$\mu(0, 0, 1) + (1 - \mu) (\tfrac{1}{2}, 0, \tfrac{1}{2}), \qquad 0 \leq \mu \leq 1.$$

1.13   Matrix games and linear programming

In this section, we formulate the general matrix game problem as a problem of linear programming. Let

$$A = (a_{ij}) = \begin{bmatrix} a_{11} & a_{12} & \cdots & a_{1n} \\ a_{21} & a_{22} & \cdots & a_{2n} \\ \cdots\cdots\cdots\cdots\cdots\cdots \\ a_{m1} & a_{m2} & \cdots & a_{mn} \end{bmatrix}$$

be the payoff matrix of a matrix game. We may assume that $a_{ij} > 0$ for all i and all j, then the value v of the game must be a positive number.

Player 1, by choosing a mixed strategy $X \in S_m$, can get at least the expected payoff

$$\min_{1 \leq j \leq n} XA_{.j} = u.$$

We have

$$XA_{.j} \geqslant u, \qquad j = 1, \ldots, n,$$

i.e.

$$\sum_{i=1}^{m} a_{ij} x_i \geqslant u, \qquad j = 1, \ldots, n,$$

$$\sum_{i=1}^{m} x_i = 1,$$

$$x_i \geqslant 0, \qquad i = 1, \ldots, m.$$

Let $x_i/u = x_i'$, $i = 1, \ldots, m$; the problem becomes

$$\sum_{i=1}^{m} a_{ij} x_i' \geqslant 1, \qquad j = 1, \ldots, n,$$

$$\sum_{i=1}^{m} x_i' = 1/u,$$

$$x_i' \geqslant 0, \qquad i = 1, \ldots, m.$$

Player 1 wishes to maximize u (by eqn (1.29), this maximum is the value v of the game); that is, he wishes to minimize $1/u$. Hence the problem reduces to the following linear programming problem:

$$\text{Minimize} \quad x_1' + \cdots + x_m',$$

subject to

$$\sum_{i=1}^{m} a_{ij} x_i' \geqslant 1, \qquad j = 1, \ldots, n,$$

$$x_i' \geqslant 0, \qquad i = 1, \ldots, m.$$

Similarly, player 2, by choosing a mixed strategy $Y \in S_n$, can keep player 1 from getting more than

$$\max_{1 \leqslant i \leqslant m} A_{i.} Y^t = w.$$

We have

$$A_{i.} Y^t \leqslant w, \qquad i = 1, \ldots, m,$$

i.e.

$$\sum_{j=1}^{n} a_{ij} y_j \leqslant w, \qquad i = 1, \ldots, m,$$

$$\sum_{j=1}^{n} y_j = 1,$$

$$y_j \geqslant 0, \qquad j = 1, \ldots, n.$$

Let $y_j/w = y_j'$, $j = 1, \ldots, n$. Since player 2 wishes to minimize w (by eqn (1.30), this minimum is also the value v of the game), that is, he wishes to maximize $1/w$, the problem reduces to the following linear programming problem, which is the dual of that formulated above:

$$\text{Maximize} \quad y_1' + \cdots + y_n',$$

subject to

$$\sum_{j=1}^{n} a_{ij} y_j' \leq 1, \qquad i = 1, \ldots, m,$$

$$y_j' \geq 0, \qquad j = 1, \ldots, n.$$

Thus the solution of a matrix game is equivalent to the problem of solving a pair of dual linear programming problems.

2

CONTINUOUS GAMES

2.1 Zero-sum two-person infinite games

A first generalization of the concept of a game is to replace each player's finite strategy set by an infinite set, e.g. by the real numbers in the closed interval $[0, 1]$.

Player 1 chooses a number x from the interval $[0, 1]$, and player 2 independently chooses a number y from the interval $[0, 1]$. x and y are called pure strategies of players 1 and 2, respectively. The choices x and y determine a play of the game and the outcome is represented by the value of a payoff function $P(x, y)$. Player 1 receives a payoff $P(x, y)$; player 2's payment is $-P(x, y)$. In other words, player 2 pays an amount $P(x, y)$ to player 1.

Such a game is called an *infinite game*. Since the sum of the payments to players 1 and 2 is always zero, this infinite game is also a zero-sum two-person game.

For example, players 1 and 2 independently choose a number x and a number y from $[0, 1]$ respectively, the payoff function being

$$P(x, y) = (x - y)^2.$$

This game is a zero-sum two-person infinite game on the unit square $0 \leqslant x \leqslant 1$, $0 \leqslant y \leqslant 1$.

If player 1 chooses $x \in [0, 1]$, his payoff is at least

$$\min_{0 \leqslant y \leqslant 1} P(x, y). \tag{2.1}$$

Since player 1 wishes to maximize his payoff, he will choose x so as to make the minimum a maximum. That is to say, player 1's payoff cannot be less than

$$\max_{0 \leqslant x \leqslant 1} \min_{0 \leqslant y \leqslant 1} P(x, y) \tag{2.2}$$

no matter which strategy player 2 chooses.

Similarly, for every fixed $y \in [0, 1]$ chosen by player 2, his loss is at most

$$\max_{0 \leqslant x \leqslant 1} P(x, y).$$

Since player 2 wishes to minimize the payoff, he will choose y so as to make the maximum a minimum. That is to say, he can prevent player 1 from getting more than

$$\min_{0\leqslant y\leqslant 1} \ \max_{0\leqslant x\leqslant 1} P(x, y). \tag{2.3}$$

Just as in the case of matrix games, we have

$$\max_{0\leqslant x\leqslant 1} \ \min_{0\leqslant y\leqslant 1} P(x, y) \leqslant \min_{0\leqslant y\leqslant 1} \ \max_{0\leqslant x\leqslant 1} P(x, y), \tag{2.4}$$

provided both sides exist. The proof is also quite similar.

If it happens that

$$\max_{0\leqslant x\leqslant 1} \ \min_{0\leqslant y\leqslant 1} P(x, y) = \min_{0\leqslant y\leqslant 1} \ \max_{0\leqslant x\leqslant 1} P(x, y), \tag{2.5}$$

then there exists a point $(x^*, y^*) \in [0, 1] \times [0, 1]$ such that

$$P(x, y^*) \leqslant P(x^*, y^*) \leqslant P(x^*, y) \tag{2.6}$$

for all $x \in [0, 1]$ and all $y \in [0, 1]$. When this is the case, we say that (x^*, y^*) is a saddle point of the payoff function $P(x, y)$, or simply a saddle point of the game. The value of $P(x, y)$ at the saddle point (x^*, y^*), i.e.

$$P(x^*, y^*), \tag{2.7}$$

is said to be the value of the game. We have

$$\max_{0\leqslant x\leqslant 1} \ \min_{0\leqslant y\leqslant 1} P(x, y) = P(x^*, y^*) = v = \min_{0\leqslant y\leqslant 1} \ \max_{0\leqslant x\leqslant 1} P(x, y),$$

$$\max_{0\leqslant x\leqslant 1} P(x, y^*) = P(x^*, y^*) = \min_{0\leqslant y\leqslant 1} P(x^*, y). \tag{2.8}$$

2.2 Mixed strategies

Now consider the case in which

$$\max_{0\leqslant x\leqslant 1} \ \min_{0\leqslant y\leqslant 1} P(x, y) < \min_{0\leqslant y\leqslant 1} \ \max_{0\leqslant x\leqslant 1} P(x, y). \tag{2.9}$$

In this case, it is necessary to introduce the notion of mixed strategies. A *mixed strategy* of player 1 is a *distribution function* $F(x)$ defined on the closed interval $[0, 1]$: for every $x \in [0, 1]$, $F(x)$ is a random process for choosing a number not greater than x. That is, $F(x)$ is the probability that the number chosen by the random process is at most x. We have

$$F(x) = \mathrm{pr}\{\xi \leqslant x\}, \tag{2.10}$$

where ξ is the random variable of the process. When $x = 0$, we define $F(0) = 0$, i.e.

$$F(0) = \mathrm{pr}\{\xi < 0\} = 0. \tag{2.11}$$

By definition,

$$F(b) - F(a) = \text{pr}\{a < \xi \leq b\}, \tag{2.12}$$

$$F(b) - F(0) = \text{pr}\{0 \leq \xi \leq b\}. \tag{2.13}$$

Every distribution function $F(x)$ defined by (2.10) and (2.11) has the following properties:

(1) $F(x)$ is non-negative, i.e.

$$F(x) \geq 0, \qquad 0 \leq x \leq 1.$$

(2) $F(0) = 0$, $F(1) = 1$.

(3) $F(x)$ is non-decreasing, i.e. if $x_1, x_2 \in [0, 1]$ and $x_1 < x_2$,

$$F(x_1) \leq F(x_2).$$

(4) $F(x)$ is a right continuous function in the open interval $(0, 1)$, i.e.

$$F(x_0 + 0) = F(x_0), \qquad 0 < x_0 < 1.$$

Property (4) can be proved as follows. Let $x_0 \in (0, 1)$, $\delta < 0$. Then it follows from the definition of a distribution function that

$$F(x_0 + \delta) - F(x_0) = \text{pr}\{x_0 < \xi \leq x_0 + \delta\}.$$

As $\delta \to 0$, the set of points x satisfying $x_0 < x \leq x_0 + \delta$ approaches the empty set as its limit. Hence

$$\text{pr}\{x_0 < \xi \leq x_0 + \delta\} \to 0,$$

so that

$$F(x_0 + \delta) \to F(x_0).$$

This proves that $F(x)$ is right continuous at any point in $(0, 1)$.

From the theory of distribution functions we know that any function $F(x)$ which satisfies the above four conditions is a distribution function.

We now return to the mixed strategies of an infinite game. When the payoff function $P(x, y)$ of a game satisfies the inequality (2.9), players 1 and 2 choose strategies x and y from $[0, 1]$ using the distribution functions $F(x)$ and $G(y)$ respectively. $F(x)$ and $G(y)$ respectively are mixed strategies of players 1 and 2.

If player 1 chooses the pure strategy x and player 2 chooses the mixed strategy $G(y)$, the expected payoff to player 1 is

$$\int_0^1 P(x, y) \, dG(y).$$

Here the integral is a Stieltjes integral.

Similarly, if player 2 uses the pure strategy y and player 1 uses the mixed strategy $F(x)$, the expected payoff to player 1 is

$$\int_0^1 P(x, y)\, \mathrm{d}F(x).$$

If players 1 and 2 use the mixed strategies $F(x)$ and $G(y)$ respectively, the expected payoff to player 1 is

$$E(F, G) = \int_0^1 \int_0^1 P(x, y)\, \mathrm{d}F(x)\, \mathrm{d}G(y). \tag{2.14}$$

Player 1 wishes to maximize this expected payoff. If he chooses a mixed strategy $F(x)$, his expected payoff is at least

$$\min_G E(F, G).$$

Player 1 can choose $F(x)$ so as to make the above minimum as large as possible. That is to say, he can be sure of an expected payoff not less than

$$v_1 = \max_F \min_G E(F, G). \tag{2.15}$$

Here the minimum and maximum are taken over the set of all distribution functions.

Similarly, player 2 can prevent player 1 from getting more than

$$v_2 = \min_G \max_F E(F, G). \tag{2.16}$$

We assume that both v_1 and v_2 exist. It is easy to prove that

$$v_1 = \max_F \min_G E(F, G) \leqslant \min_G \max_F E(F, G) = v_2. \tag{2.17}$$

The proof is similar to that of Theorem 1.2.

2.3　Continuous games

In general, the two quantities v_1 and v_2 in (2.17) might not be equal. We now state without proof the following fundamental theorem for continuous games. For the proof the reader is referred to McKinsey (1952) or Vorob'ev (1977).

Theorem 2.1. *If the payoff function of an infinite game is continuous on $0 \leqslant x \leqslant 1$, $0 \leqslant y \leqslant 1$, then*

$$v_1 = \max_F \min_G \int_0^1 \int_0^1 P(x, y)\, \mathrm{d}F(x)\, \mathrm{d}G(y)$$

and

$$v_2 = \min_G \max_F \int_0^1 \int_0^1 P(x, y)\, \mathrm{d}F(x)\, \mathrm{d}G(y)$$

exist and are equal to each other.

An infinite game with a continuous payoff function is called a *continuous game*. In the remaining part of this chapter, we shall deal only with continuous games on the unit square $0 \leq x \leq 1$, $0 \leq y \leq 1$, unless otherwise stated.

When $v_1 = v_2$, the common value $v = v_1 = v_2$ is called the value of the game. In this case there exist optimal mixed strategies $F^*(x)$, $G^*(y)$ such that

$$E(F, G^*) \leq E(F^*, G^*) \leq E(F^*, G) \tag{2.18}$$

for all distribution functions F and G. As in the case of matrix games, (F^*, G^*) is called a saddle point of $E(F, G)$, or simply a saddle point of the continuous game. It is also called a solution of the game.

2.4   Properties of optimal strategies

We shall first prove a property of the integral of a continuous function with respect to a distribution function.

Theorem 2.2. *Let* $f(x)$, $g(y)$ *be continuous functions on the closed interval* $[0, 1]$, *and let* $F(x)$, $G(y)$ *be distribution functions. Then*

$$\max_F \int_0^1 f(x)\, \mathrm{d}F(x) = \max_{0 \leq x \leq 1} f(x), \tag{2.19}$$

$$\min_G \int_0^1 g(y)\, \mathrm{d}G(y) = \min_{0 \leq y \leq 1} g(y). \tag{2.20}$$

Proof. Suppose that

$$\max_{0 \leq x \leq 1} f(x) = f(a). \tag{2.21}$$

Then

$$f(a) \geq f(x), \qquad 0 \leq x \leq 1.$$

For every distribution function $F(x)$ we have

$$f(a) = \int_0^1 f(a)\, \mathrm{d}F(x) \geq \int_0^1 f(x)\, \mathrm{d}F(x).$$

Hence

$$f(a) \geqslant \sup_F \int_0^1 f(x)\, dF(x). \qquad (2.22)$$

Now consider the special distribution function, namely the step function

$$I_a(x) = \begin{cases} 0, & 0 \leqslant x < a, \\ 1, & a \leqslant x \leqslant 1. \end{cases}$$

We have

$$\sup_F \int_0^1 f(x)\, dF(x) \geqslant \int_0^1 f(x)\, dI_a(x). \qquad (2.23)$$

It is easily verified that

$$\int_0^1 f(x)\, dI_a(x) = f(a). \qquad (2.24)$$

It follows from (2.22), (2.23), (2.24) that

$$\sup_F \int_0^1 f(x)\, dF(x) = \int_0^1 f(x)\, dI_a(x) = f(a). \qquad (2.25)$$

That is to say,

$$\int_0^1 f(x)\, dF(x)$$

assumes its supremum with respect to F at $F(x) = I_a(x)$, and the value of the supremum is $f(a)$, i.e.

$$\sup_F \int_0^1 f(x)\, dF(x) = \max_F \int_0^1 f(x)\, dF(x). \qquad (2.26)$$

From (2.26), (2.25) and (2.21) we obtain

$$\max_F \int_0^1 f(x)\, dF(x) = \max_{0 \leqslant x \leqslant 1} f(x).$$

In a similar manner it can be proved that

$$\min_G \int_0^1 g(y)\, dG(y) = \min_{0 \leqslant y \leqslant 1} g(y). \qquad \square$$

This theorem states that the minimum (maximum) of a continuous function with respect to mixed strategies is equal to the minimum (maximum) of that function. (2.19) and (2.20) can be viewed as generalizations of relations (1.27) and (1.28).

Using Theorem 2.2 we can write Theorem 2.1 in another form. Since

$$\max_F \int_0^1 f(x) \, dF(x) = \max_{0 \leqslant x \leqslant 1} f(x),$$

we have

$$\max_F \int_0^1 \left[\int_0^1 P(x, y) \, dG(y) \right] dF(x) = \max_{0 \leqslant x \leqslant 1} \int_0^1 P(x, y) \, dG(y). \quad (2.27)$$

Hence

$$\min_G \max_F \int_0^1 \int_0^1 P(x, y) \, dF(x) \, dG(y) = \max_F \int_0^1 \int_0^1 P(x, y) \, dG^*(y) \, dF(x)$$

$$= \max_{0 \leqslant x \leqslant 1} \int_0^1 P(x, y) \, dG^*(y). \quad (2.28)$$

Similarly,

$$\max_F \min_G \int_0^1 \int_0^1 P(x, y) \, dF(x) \, dG(y) = \min_G \int_0^1 \int_0^1 P(x, y) \, dF^*(x) \, dG(y)$$

$$= \min_{0 \leqslant y \leqslant 1} \int_0^1 P(x, y) \, dF^*(x). \quad (2.29)$$

Therefore, we can rewrite the fundamental theorem of continuous games (Theorem 2.1) in the following form:

$$\max_{0 \leqslant x \leqslant 1} \int_0^1 P(x, y) \, dG^*(y) = E(F^*, G^*) = \min_{0 \leqslant y \leqslant 1} \int_0^1 P(x, y) \, dF^*(x), \quad (2.30)$$

where $F^*(x)$, $G^*(y)$ are respectively optimal mixed strategies of players 1 and 2.

The following theorem concerning continuous games is the analogue of Theorem 1.6 for matrix games.

Theorem 2.3. *Let $P(x, y)$ by the payoff function of a continuous game whose value is v.*

(1) *Let $G^*(y)$ be an optimal strategy of player 2. If*

$$\int_0^1 P(x_0, y) \, dG^*(y) < v$$

for $x_0 \in [0, 1]$, then

$$\text{pr}\{\xi = x_0\} = 0.$$

(2) *Let $F^*(x)$ be an optimal strategy of player 1. If*

$$\int_0^1 P(x, y_0) \, dF^*(x) > v$$

for $y_0 \in [0, 1]$, then

$$\mathrm{pr}\{\eta = y_0\} = 0.$$

Proof. Only (1) will proved. The proof of (2) is similar. Since $G^*(y)$ is an optimal strategy of player 2, we have

$$\int_0^1 P(x, y) \, dG^*(y) \leqslant v.$$

Let

$$S_1 = \left\{ x: \int_0^1 P(x, y) \, dG^*(y) < v \right\},$$

$$S_2 = \left\{ x: \int_0^1 P(x, y) \, dG^*(y) = v \right\},$$

and let $F^*(x)$ be an optimal strategy of player 1. Then

$$v = \int_0^1 \left[\int_0^1 P(x, y) \, dG^*(y) \right] dF^*(x)$$

$$= \int_{S_1} \left[\int_0^1 P(x, y) \, dG^*(y) \right] dF^*(x) + \int_{S_2} \left[\int_0^1 P(x, y) \, dG^*(y) \right] dF^*(x)$$

$$= \int_{S_1} \left[\int_0^1 P(x, y) \, dG^*(y) \right] dF^*(x) + v \int_{S_2} dF^*(x).$$

Hence

$$v \left[1 - \int_{S_2} dF^*(x) \right] = \int_{S_1} \left[\int_0^1 P(x, y) \, dG^*(y) \right] dF^*(x),$$

i.e.

$$v \int_{S_1} dF^*(x) = \int_{S_1} \left[\int_0^1 P(x, y) \, dG^*(y) \right] dF^*(x),$$

or

$$\int_{S_1} \left[v - \int_0^1 P(x, y) \, dG^*(y) \right] dF^*(x) = 0.$$

When $x = x_0 \in S_1$,

$$v - \int_0^1 P(x, y) \, dG^*(y) > 0.$$

In the integral in the last equation above, the infinitesimal part corresponding to x_0 is

$$\left[v - \int_0^1 P(x, y)\, dG^*(y)\right]\left[F^*(x_0) - F^*(x_0 - 0)\right] = 0.$$

Hence we must have

$$\mathrm{pr}\{\xi = x_0\} = F^*(x_0) - F^*(x_0 - 0) = 0. \qquad \square$$

Corresponding to Theorem 1.7, we have the following theorem for continuous games.

Theorem 2.4. *Let* $P(x, y)$ *be the payoff function of a continuous game whose value is* v.

(1) *A necessary and sufficient condition for the distribution function* $F^*(x)$ *to be an optimal strategy of player 1 is*

$$v \leqslant \int_0^1 P(x, y)\, dF^*(x), \qquad 0 \leqslant y \leqslant 1.$$

(2) *A necessary and sufficient condition for the distribution function* $G^*(y)$ *to be an optimal strategy of player 2 is*

$$\int_0^1 P(x, y)\, dG^*(y) \leqslant v, \qquad 0 \leqslant x \leqslant 1.$$

Proof. Only (1) will be proved. The proof of (2) is similar.
  Necessity. By (2.30)

$$v = \min_{0 \leqslant y \leqslant 1} \int_0^1 P(x, y)\, dF^*(x).$$

Hence

$$v \leqslant \int_0^1 P(x, y)\, dF^*(x), \qquad 0 \leqslant y \leqslant 1.$$

Sufficiency. Assume that

$$v \leqslant \int_0^1 P(x, y)\, dF^*(x), \qquad 0 \leqslant y \leqslant 1. \qquad (2.31)$$

Let (F^0, G^0) be a saddle point of the game, i.e.

$$E(F, G^0) \leqslant E(F^0, G^0) \leqslant E(F^0, G) \qquad (2.32)$$

for all distribution functions F and G.
  We will prove that (F^*, G^0) is a saddle point of the game. Integrate

both sides of (2.31) with respect to any distribution function $G(y)$ to obtain

$$\int_0^1 v \, dG(y) = v \leqslant E(F^*, G). \tag{2.33}$$

In particular,

$$v \leqslant E(F^*, G^0). \tag{2.34}$$

The definition of saddle point (2.32) implies

$$E(F^*, G^0) \leqslant E(F^0, G^0) = v. \tag{2.35}$$

From (2.34) and (2.35) we have

$$E(F^*, G^0) = E(F^0, G^0) = v. \tag{2.36}$$

It follows from (2.32), (2.36) and (2.33) that

$$E(F, G^0) \leqslant E(F^*, G^0) \leqslant E(F^*, G),$$

which proves that (F^*, G^0) is a saddle point of the game. Hence, $F^*(x)$ is an optimal strategy of player 1. $\qquad \square$

Example 2.1. Let the payoff function of a continuous game be

$$P(x, y) = (x - y)^2, \qquad 0 \leqslant x \leqslant 1, \qquad 0 \leqslant y \leqslant 1.$$

In this game, the value of the game is $v = \frac{1}{4}$. The optimal strategy of player 2 is the pure strategy $y = \frac{1}{2}$. The optimal strategy of player 1 is to choose $x = 0$ and $x = 1$ with equal probability. That is,

$$G^*(y) = I_{\frac{1}{2}}(y), \qquad F^*(x) = \tfrac{1}{2} I_0(x) + \tfrac{1}{2} I_1(x).$$

Let us verify these facts. (See Section 2.5 below.) First, we have

$$\max_{0 \leqslant x \leqslant 1} \int_0^1 P(x, y) \, dG^*(y) = \max_{0 \leqslant x \leqslant 1} \int_0^1 (x - y)^2 \, dI_{\frac{1}{2}}(y)$$

$$= \max_{0 \leqslant x \leqslant 1} (x - \tfrac{1}{2})^2 = \tfrac{1}{4}.$$

Next,

$$\min_{0 \leqslant y \leqslant 1} \int_0^1 P(x, y) \, dF^*(x) = \min_{0 \leqslant y \leqslant 1} \int_0^1 (x - y)^2 \, d[\tfrac{1}{2} I_0(x) + \tfrac{1}{2} I_1(x)]$$

$$= \min_{0 \leqslant y \leqslant 1} [\tfrac{1}{2}(0 - y)^2 + \tfrac{1}{2}(1 - y)^2]$$

$$= \tfrac{1}{4}.$$

We obtain

$$\max_{0 \leqslant x \leqslant 1} \int_0^1 (x-y)^2 \, dG^*(y) = \tfrac{1}{4} = \min_{0 \leqslant y \leqslant 1} \int_0^1 (x-y)^2 \, dF^*(x).$$

Hence, by (2.30),

$$G^*(y) = I_{\frac{1}{2}}(y)$$

and

$$F^*(x) = \tfrac{1}{2}I_0(x) + \tfrac{1}{2}I_1(x)$$

are optimal strategies of players 1 and 2, respectively, and $v = \tfrac{1}{4}$ is the value of the game.

2.5 Convex games

Let us first introduce the definition of a convex function.

Definition 2.1. Let $f(y)$ be a function defined on $[0, 1]$. If for every pair of points $y_1, y_2 \in [0, 1]$ and every λ, $0 \leqslant \lambda \leqslant 1$, we have

$$f(\lambda y_1 + (1 - \lambda)y_2) \leqslant \lambda f(y_1) + (1 - \lambda)f(y_2), \qquad (2.37)$$

then $f(y)$ is called a *convex function* on $[0, 1]$.

If the strict inequality holds in (2.37) for all λ, $0 < \lambda < 1$, then $f(y)$ is called a *strictly convex function*.

The geometrical interpretation of a strictly convex function is as follows. The line segment connecting any two points of the graph of a strictly convex function always lies above the graph of the function between the two points. See Fig. 2.1.

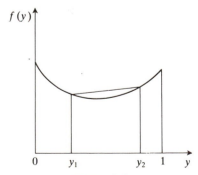

Figure 2.1

If the payoff function of a continuous game is convex in one variable, the game is called a *convex game*, or a game with convex payoff function.

Let $P(x, y)$ be the payoff function of a continuous game on the unit square. We assume that, for each x, $P(x, y)$ is strictly convex in y. We shall determine in this section the solution of this class of continuous games.

Lemma. *Let the function $P(x, y)$ be continuous on the unit square $0 \leqslant x \leqslant 1$, $0 \leqslant y \leqslant 1$. If for each $x \in [0, 1]$, $P(x, y)$ is strictly convex in y, then the function $\phi(y)$ defined by the integral*

$$\phi(y) = \int_0^1 P(x, y) \, \mathrm{d}F(x),$$

where $F(x)$ is any distribution function, is a continuous function of y and is strictly convex in y.

Proof. Let us first prove the continuity of $\phi(y)$. Since $P(x, y)$ is continuous in y, for any $\varepsilon > 0$, there exists $\delta > 0$ such that

$$|P(x, y_1) - P(x, y_2)| < \varepsilon$$

whenever $|y_1 - y_2| < \delta$. Hence

$$|\phi(y_1) - \phi(y_2)| = \left| \int_0^1 P(x, y_1) \, \mathrm{d}F(x) - \int_0^1 P(x, y_2) \, \mathrm{d}F(x) \right|$$

$$\leqslant \int_0^1 |P(x, y_1) - P(x, y_2)| \, \mathrm{d}F(x)$$

$$< \varepsilon \int_0^1 \mathrm{d}F(x) = \varepsilon.$$

The continuity of $\phi(y)$ is proved.

Now let us prove that $\phi(y)$ is strictly convex in y. Let $0 < \lambda < 1$. Then

$$\phi(\lambda y_1 + (1 - \lambda) y_2) = \int_0^1 P(x, \lambda y_1 + (1 - \lambda) y_2) \, \mathrm{d}F(x)$$

$$< \int_0^1 [\lambda P(x, y_1) + (1 - \lambda) P(x, y_2)] \, \mathrm{d}F(x)$$

$$= \lambda \int_0^1 P(x, y_1) \, \mathrm{d}F(x) + (1 - \lambda) \int_0^1 P(x, y_2) \, \mathrm{d}F(x)$$

$$= \lambda \phi(y_1) + (1 - \lambda) \phi(y_2).$$

Hence $\phi(y)$ is a strictly convex function of y. $\qquad \square$

We now proceed to discuss the solution of convex games.

Theorem 2.5. *Let $P(x, y)$ be the payoff function of a continuous game on the unit square, and let $P(x, y)$ be strictly convex in y for each x. Then the optimal strategy of player 2 is a unique pure strategy.*

Proof. By Theorem 2.1 solution of the game exists. We will show that the only optimal strategy for player 2 is a pure strategy. For this purpose, let $F^*(x)$ and $G^*(y)$ be any optimal strategies of players 1 and 2, respectively. We need only show that

$$G^*(y) = I_{y^*}(y),$$

i.e.

$$G^*(y) = \begin{cases} 0, & 0 \leqslant y < y^*, \\ 1, & y^* \leqslant y \leqslant 1. \end{cases}$$

This is equivalent to saying that for every $\varepsilon > 0$ we have

$$G^*(y^* + \varepsilon) - G^*(y^* - \varepsilon) = 1. \tag{2.38}$$

The proof of this is given below. Let

$$\phi(y) = \int_0^1 P(x, y)\, dF^*(x).$$

Then $\phi(y)$ is continuous and strictly convex, and hence it assumes its minimum at a unique point $y = y^* \in [0, 1]$. Thus for every $\varepsilon > 0$, there exists a $\delta > 0$, such that

$$\begin{aligned} \phi(y) &\geqslant \phi(y^*) && \text{for } y \in [y^* - \varepsilon, y^* + \varepsilon], \\ \phi(y) &\geqslant \phi(y^*) + \delta && \text{for } y \in [0, y^* - \varepsilon], [y^* + \varepsilon, 1]. \end{aligned}$$

We have

$$\begin{aligned} v &= \int_0^1 \phi(y)\, dG^*(y) = \left(\int_0^{y^*-\varepsilon} + \int_{y^*-\varepsilon}^{y^*+\varepsilon} + \int_{y^*+\varepsilon}^1 \right) \phi(y)\, dG^*(y) \\ &\geqslant \int_0^{y^*-\varepsilon} [\phi(y^*) + \delta]\, dG^*(y) + \int_{y^*-\varepsilon}^{y^*+\varepsilon} \phi(y^*)\, dG^*(y) \\ &\quad + \int_{y^*+\varepsilon}^1 [\phi(y^*) + \delta]\, dG^*(y) \\ &= \int_0^1 \phi(y^*)\, dG^*(y) + \delta[G^*(y^* - \varepsilon) - 0 + 1 - G^*(y^* + \varepsilon)] \\ &= \phi(y^*) + \delta[1 - G^*(y^* + \varepsilon) + G^*(y^* - \varepsilon)]. \end{aligned} \tag{2.39}$$

But

$$\phi(y^*) = \int_0^1 P(x, y^*)\, \mathrm{d}F^*(x)$$

$$= \min_{0 \leqslant y \leqslant 1} \int_0^1 P(x, y)\, \mathrm{d}F^*(x) = v$$

by (2.30). Hence (2.39) yields

$$v \geqslant v + \delta[1 - G^*(y^* + \varepsilon) + G^*(y^* - \varepsilon)],$$

or

$$G^*(y^* + \varepsilon) - G^*(y^* - \varepsilon) \geqslant 1.$$

Now by the definition of a distribution function,

$$G^*(y^* + \varepsilon) - G^*(y^* - \varepsilon) \leqslant 1.$$

Therefore,

$$G^*(y^* + \varepsilon) - G^*(y^* - \varepsilon) = 1. \qquad \square$$

By means of this theorem we can find the value v and the optimal pure strategy y^* of player 2 for the convex game. We have

$$v = \min_{G} \max_{F} \int_0^1 \left[\int_0^1 P(x, y)\, \mathrm{d}G(y) \right] \mathrm{d}F(x).$$

Using eqn (2.27) the above expression can be written as

$$v = \min_{G} \max_{0 \leqslant x \leqslant 1} \int_0^1 P(x, y)\, \mathrm{d}G(y).$$

It is known that the optimal strategy of player 2 is a pure strategy. Hence in taking the minimum we need consider only those distribution functions $G(y)$ which are step functions. Therefore,

$$v = \min_{0 \leqslant y \leqslant 1} \max_{0 \leqslant x \leqslant 1} \int_0^1 P(x, y)\, \mathrm{d}I_y(y)$$

$$= \min_{0 \leqslant y \leqslant 1} \max_{0 \leqslant x \leqslant 1} P(x, y).$$

Thus we have the following theorem.

Theorem 2.6. *Under the hypotheses of Theorem 2.5, the value of the convex game is*

$$v = \min_{0 \leqslant y \leqslant 1} \max_{0 \leqslant x \leqslant 1} P(x, y). \qquad (2.40)$$

It follows from the above discussion that in a convex game player 2 has a unique optimal strategy which is a pure strategy y^* satisfying

$$v = \min_{0 \leqslant y \leqslant 1} \max_{0 \leqslant x \leqslant 1} P(x, y) = \max_{0 \leqslant x \leqslant 1} P(x, y^*). \qquad (2.41)$$

We now turn to the optimal strategies of player 1. We have the following three cases.

(1) $y^* = 0$. By (2.41),

$$\max_{0 \leqslant x \leqslant 1} P(x, 0) = v. \qquad (2.42)$$

Hence, for every $x \in [0, 1]$ we have

$$P(x, 0) \leqslant v.$$

Let

$$S_0 = \{x_0 : P(x_0, 0) = v\}, \qquad S_1 = \{x_1 : P(x_1, 0) < v\}.$$

Then $S_0 \cup S_1 = [0, 1]$. We shall show that there exists an optimal pure strategy for player 1. In view of eqn (2.8) this is equivalent to saying that there exists $x_0 \in S_0$ such that

$$\min_{0 \leqslant y \leqslant 1} P(x_0, y) = v = P(x_0, 0),$$

i.e. $P(x_0, y)$ is non-decreasing at $y = 0$.

Suppose to the contrary that for every $x_0 \in S_0$, $P(x_0, y)$ is monotone decreasing at $y = 0$. Then in every neighborhood of $y = 0$ there are values of $P(x_0, y)$ smaller than $P(x_0, 0)$. That is to say, for every $x_0 \in S_0$, there exists $\delta > 0$ such that

$$P(x_0, y) < P(x_0, 0) = v \qquad (2.43)$$

for $0 < y < \delta$. Moreover, since $P(x_1, y)$ is continuous at $y = 0$, then for every $x_1 \in S_1$ there exists $\delta > 0$ such that

$$P(x_1, y) < v \qquad (2.44)$$

for $0 < y < \delta$.

Combining (2.43) and (2.44): for every $x \in [0, 1]$, there exists $\delta > 0$ such that

$$P(x, y) < v \qquad (2.45)$$

for $0 < y < \delta$.

For every x, define $\delta(x)$ to be the supremum of all numbers δ satisfying (2.45). It follows from the continuity of P that $\delta(x)$ is a continuous function of x on the closed interval $[0, 1]$. Since $\delta(x)$ is always positive, its minimum is also positive. Let $\delta_0 > 0$ be this minimum. Let y_1 be a number satisfying $0 < y_1 < \delta_0$. Then

$$\max_{0 \leqslant x \leqslant 1} P(x, y_1) < v.$$

From (2.40) we have

$$v = \min_{0 \leqslant y \leqslant 1} \max_{0 \leqslant x \leqslant 1} P(x, y) = \max_{0 \leqslant x \leqslant 1} P(x, y_1) < v,$$

a contradiction. Hence we conclude that there exists $x_0 \in S_0$ such that $P(x_0, 0) = v$ and such that $P(x_0, y)$ is non-decreasing at $y = 0$.

If the partial derivative $\partial P(x, y)/\partial y$ exists, then the above conclusion is equivalent to: there exists player 1's optimal pure strategy $x^* \in [0, 1]$ satisfying

$$P(x^*, 0) = v,$$

$$\frac{\partial}{\partial y} P(x^*, 0) \geqslant 0. \tag{2.46}$$

Here the partial derivative of P with respect to y at $y = 0$ should be understood as the right-hand derivative.

(2) $y^* = 1$. In a similar manner we can prove that there exists player 1's optimal pure strategy $x^* \in [0, 1]$ satisfying

$$P(x^*, 1) = v,$$

$$\frac{\partial}{\partial y} P(x^*, 1) \leqslant 0. \tag{2.47}$$

Here the partial derivative of P should be understood as the left-hand derivative.

(3) $0 < y^* < 1$. As in the case (1), we have

$$P(x, y^*) \leqslant v \tag{2.48}$$

for all $x \in [0, 1]$, and

$$P(x, y^*) = v \tag{2.49}$$

for some $x \in [0, 1]$.

If every x satisfying (2.49) also satisfies

$$\frac{\partial}{\partial y} P(x, y^*) < 0, \tag{2.50}$$

we would be led to the same contradiction as in (1). Therefore, there exists $x_1^* \in [0, 1]$ such that

$$P(x_1^*, y^*) = v,$$

$$\frac{\partial}{\partial y} P(x_1^*, y^*) \geqslant 0. \tag{2.51}$$

By an analogous argument, we see that there exists $x_2^* \in [0, 1]$ such that

$$P(x_2^*, y^*) = v,$$

$$\frac{\partial}{\partial y} P(x_2^*, y^*) \leqslant 0. \tag{2.52}$$

Let us now consider the function

$$f(\lambda) = \lambda \frac{\partial}{\partial y} P(x_1^*, y^*) + (1 - \lambda) \frac{\partial}{\partial y} P(x_2^*, y^*).$$

This is a linear function of λ. We have

$$f(0) = \frac{\partial}{\partial y} P(x_2^*, y^*) \leqslant 0, \qquad f(1) = \frac{\partial}{\partial y} P(x_1^*, y^*) \geqslant 0.$$

Hence there must be a number α, $0 \leqslant \alpha \leqslant 1$, such that $f(\alpha) = 0$. That is,

$$f(\alpha) = \alpha \frac{\partial}{\partial y} P(x_1^*, y^*) + (1 - \alpha) \frac{\partial}{\partial y} P(x_2^*, y^*) = 0. \tag{2.53}$$

We will prove that if $x_1^* \in [0, 1]$, $x_2^* \in [0, 1]$ and $\alpha \in [0, 1]$ are three numbers satisfying (2.51), (2.52), (2.53), then the distribution function

$$F^*(x) = \alpha I_{x_1^*}(x) + (1 - \alpha) I_{x_2^*}(x)$$

is an optimal strategy for player 1.

In the first place, the function

$$g(y) = \alpha P(x_1^*, y) + (1 - \alpha) P(x_2^*, y)$$

is a strictly convex function. Using (2.53) we see that the derivative of $g(y)$ vanishes at $y = y^*$:

$$g'(y^*) = f(\alpha) = 0.$$

Hence $g(y)$ assumes its minimum at y^*, the minimum value being

$$g(y^*) = \alpha P(x_1^*, y^*) + (1 - \alpha) P(x_2^*, y^*)$$
$$= \alpha v + (1 - \alpha) v = v.$$

We have proved that for all $y \in [0, 1]$,

$$v = g(y^*) \leqslant g(y) = \alpha P(x_1^*, y) + (1 - \alpha) P(x_2^*, y).$$

That is,

$$v \leqslant \int_0^1 P(x, y) \, \mathrm{d}[\alpha I_{x_1^*}(x) + (1 - \alpha) I_{x_2^*}(x)].$$

By Theorem 2.4(1), this proves that the distribution function

$$F^*(x) = \alpha I_{x_1^*}(x) + (1 - \alpha)I_{x_2^*}(x)$$

is an optimal strategy for player 1.

We now summarize the results of the cases (1), (2), (3) in the following theorem.

Theorem 2.7. *Under the hypotheses of Theorem 2.5, player 1 has an optimal strategy $F^*(x)$.*
(1) *If $y^* = 0$, then*

$$F^*(x) = I_{x^*}(x), \tag{2.54}$$

where $x^* \in [0, 1]$ satisfies the following conditions:

$$P(x^*, 0) = v,$$
$$\frac{\partial}{\partial y} P(x^*, 0) \geqslant 0. \tag{2.55}$$

(2) *If $y^* = 1$, then*

$$F^*(x) = I_{x^*}(x), \tag{2.56}$$

where $x^* \in [0, 1]$ satisfies the following conditions:

$$P(x^*, 1) = v,$$
$$\frac{\partial}{\partial y} P(x^*, 1) \leqslant 0. \tag{2.57}$$

(3) *If $0 < y^* < 1$, then*

$$F^*(x) = \alpha I_{x_1^*}(x) + (1 - \alpha)I_{x_2^*}(x), \qquad 0 \leqslant \alpha \leqslant 1, \tag{2.58}$$

where $x_1^* \in [0, 1]$, $x_2^* \in [0, 1]$ and α satisfy the following conditions:

$$P(x_1^*, y^*) = v,$$
$$\frac{\partial}{\partial y} P(x_1^*, y^*) \geqslant 0,$$
$$P(x_2^*, y^*) = v,$$
$$\frac{\partial}{\partial y} P(x_2^*, y^*) \leqslant 0, \tag{2.59}$$

$$\alpha \frac{\partial}{\partial y} P(x_1^*, y^*) + (1 - \alpha)\frac{\partial}{\partial y} P(x_2^*, y^*) = 0.$$

It should be noted that if the payoff function $P(x, y)$ of the game is convex instead of being strictly convex in y, the theorems in this section are still true, except that in this case the optimal strategy for player 2 may not be unique.

Example 2.2. Let the payoff function of a continuous game on the unit square $0 \leqslant x \leqslant 1$, $0 \leqslant y \leqslant 1$ be

$$P(x, y) = (x - y)^2.$$

Since

$$\frac{\partial^2}{\partial y^2} P(x, y) = 2 > 0,$$

$P(x, y)$ is strictly convex in y for each $x \in [0, 1]$.

Let v be the value of the game. By the formula (2.40),

$$v = \min_{0 \leqslant y \leqslant 1} \max_{0 \leqslant x \leqslant 1} (x - y)^2$$

$$= \min \begin{cases} (1 - y)^2, & 0 \leqslant y \leqslant \frac{1}{2}, \\ (0 - y)^2, & \frac{1}{2} \leqslant y \leqslant 1, \end{cases}$$

see Fig. 2.2. The heavy curve in the figure is the graph of

$$\max_{0 \leqslant x \leqslant 1} (x - y)^2.$$

It is easily seen that, in the interval $0 \leqslant y \leqslant \frac{1}{2}$, $(1 - y)^2$ assumes its minimal value of $\frac{1}{4}$ at $y = \frac{1}{2}$; in the interval $\frac{1}{2} \leqslant y \leqslant 1$, $(0 - y)^2$ also assumes

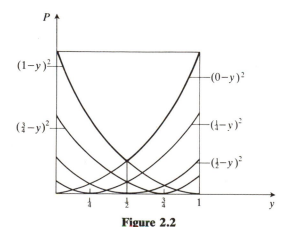

Figure 2.2

its minimal value of $\frac{1}{4}$ at $y = \frac{1}{2}$. Hence

$$v = \tfrac{1}{4} = \max_{0 \leqslant x \leqslant 1} (x - \tfrac{1}{2})^2$$

and the optimal pure strategy of player 2 is $y^* = \frac{1}{2}$.

Now let us find player 1's optimal strategy. We first solve

$$P(x, y^*) = P(x, \tfrac{1}{2}) = (x - \tfrac{1}{2})^2 = \tfrac{1}{4} = v$$

to obtain

$$x_1^* = 0, \qquad x_2^* = 1.$$

We have

$$\frac{\partial}{\partial y} P(0, \tfrac{1}{2}) = 1 \geqslant 0 \geqslant -1 = \frac{\partial}{\partial y} P(1, \tfrac{1}{2}).$$

The value of α can then be determined. By the last formula in (2.59),

$$\alpha \frac{\partial}{\partial y} P(0, \tfrac{1}{2}) + (1 - \alpha) \frac{\partial}{\partial y} P(1, \tfrac{1}{2}) = 0,$$

i.e.

$$\alpha[-2(0 - \tfrac{1}{2})] + (1 - \alpha)[-2(1 - \tfrac{1}{2})] = 0.$$

Thus $\alpha = \frac{1}{2}$. Player 1's optimal strategy is

$$F^*(x) = \tfrac{1}{2}I_0(x) + \tfrac{1}{2}I_1(x).$$

2.6 Separable games

Definition 2.2. If the payoff function of an infinite game on the unit square $0 \leqslant x \leqslant 1$, $0 \leqslant y \leqslant 1$ has the form

$$P(x, y) = \sum_{i=1}^{m} \sum_{j=1}^{n} a_{ij} r_i(x) s_j(y), \qquad (2.60)$$

where a_{ij} are constants, $r_i(x)$, $s_j(y)$ are continuous functions, $i = 1, \ldots, m$, $j = 1, \ldots, n$, the game is called a *separable game*.

The payoff function of a separable game is called a *separable function*.

A polynomial in two variables is obviously a special case of a separable function.

Let (2.60) be the payoff function of a separable game; and let players 1 and 2 use as their mixed strategies the distribution functions $F(x)$ and

$G(y)$, respectively. Then the expected payoff to player 1 is

$$\int_0^1 \int_0^1 P(x, y) \, dF(x) \, dG(y)$$

$$= \int_0^1 \int_0^1 \left[\sum_{i=1}^m \sum_{j=1}^n a_{ij} r_i(x) s_j(y) \right] dF(x) \, dG(y) \tag{2.61}$$

$$= \sum_{i=1}^m \sum_{j=1}^n a_{ij} \int_0^1 \int_0^1 r_i(x) s_j(y) \, dF(x) \, dG(y)$$

$$= \sum_{i=1}^m \sum_{j=1}^n a_{ij} \int_0^1 r_i(x) \, dF(x) \int_0^1 s_j(y) \, dG(y).$$

Denote

$$r_i = \int_0^1 r_i(x) \, dF(x), \qquad i = 1, \ldots, m, \tag{2.62}$$

$$s_j = \int_0^1 s_j(y) \, dG(y), \qquad j = 1, \ldots, n. \tag{2.63}$$

Then (2.61) can be written in the form

$$E(r, s) = \sum_{i=1}^m \sum_{j=1}^n a_{ij} r_i s_j, \tag{2.64}$$

where $r = (r_1, \ldots, r_m)$, $s = (s_1, \ldots, s_n)$.

Thus to each distribution function $F(x)$ there corresponds an r. When $F(x)$ varies over the set of all distribution functions, the set of all r is a subset R of the m-dimensional Euclidean space. Similarly, to each distribution function $G(y)$ there corresponds an s. When $G(y)$ varies over the set of all distribution functions, the set of all s is a subset S of the n-dimensional Euclidean space. The sets R and S are called the *moment spaces* of the functions $\{r_i(x)\}$ and $\{s_j(y)\}$, respectively.

Hence, in a separable game, the selection of a mixed strategy for player 1 is equivalent to the selection of a point r from the moment space R, and for player 2 it is equivalent to the selection of a point s from the moment space S. The expected payoff to player 1 is given by (2.64).

Consider

$$r_i = r_i(x), \qquad i = 1, \ldots, m, \tag{2.65}$$

where x varies from 0 to 1. Equations (2.65) are the parametric equations of a space curve C. Let H be the convex set spanned by the curve C, i.e. the convex hull of C. Similarly, let C' be the curve

$$s_j = s_j(y), \qquad j = 1, \ldots, n, \qquad 0 \leqslant y \leqslant 1, \tag{2.66}$$

and H' be the convex set spanned by C', i.e. the convex hull of C'.

The following theorem establishes the relationship between R and H (S and H').

Theorem 2.8. $R = H(S = H')$.

Proof. (1) We first prove $R \subseteq H$, i.e. $r^0 \in R$ implies $r^0 \in H$. Assume that $r^0 = (r_1^0, \ldots, r_m^0) \in R$, $r^0 \notin H$. Let $F^0(x)$ be a distribution function which generates r^0, i.e.

$$r_i^0 = \int_0^1 r_i(x) \, dF^0(x), \qquad i = 1, \ldots, m.$$

Since H is a convex set, and $r^0 \notin H$, by the theorem of the supporting hyperplanes, there exists a hyperplane

$$p: \quad \sum_{i=1}^m c_i z_i + b = 0$$

such that $r^0 \in p$ and such that H lies entirely in one of the two half-spaces formed by p. That is to say, we have

$$\sum_{i=1}^m c_i r_i^0 + b = 0 \tag{2.67}$$

and

$$\sum_{i=1}^m c_i r_i(x) + b < 0, \qquad 0 \leq x \leq 1. \tag{2.68}$$

Hence there exists $\delta > 0$ such that

$$\sum_{i=1}^m c_i r_i(x) + b \leq -\delta < 0, \qquad 0 \leq x \leq 1. \tag{2.69}$$

(The proof is similar to that of Lemma 1 in Section 1.5. The $\delta > 0$ here is the shortest distance from H to r^0 in section 16.3 of von Neumann and Morganstern (1953).)

From (2.67) and (2.69) we have

$$\sum_{i=1}^m c_i r_i^0 - \sum_{i=1}^m c_i r_i(x) \geq \delta > 0, \qquad 0 \leq x \leq 1.$$

Integrating with respect to the distribution function $F^0(x)$:

$$\sum_{i=1}^m c_i r_i^0 \int_0^1 dF^0(x) - \sum_{i=1}^m c_i \int_0^1 r_i(x) \, dF^0(x) \geq \delta \int_0^1 dF^0(x),$$

i.e.

$$\sum_{i=1}^{m} c_i r_i^0 - \sum_{i=1}^{m} c_i r_i^0 \geqslant \delta > 0.$$

We obtain $0 > 0$. This is an absurdity, hence $R \subseteq H$.

(2) We now prove $H \subseteq R$. Assume that $r^0 = (r_1^0, \ldots, r_m^0) \in H$. Then r^0 can be expressed as a convex linear combination of m points of the curve C (for the proof see Karlin (1959, vol. 2, p. 358) or McKinsey (1952, theorem 2.1)), i.e.

$$r_i^0 = \sum_{k=1}^{m} \alpha_k r_i(x_k), \quad i = 1, \ldots, m,$$

where $\alpha_k \geqslant 0$, $\sum_{k=1}^{m} \alpha_k = 1$, and $0 \leqslant x_k \leqslant 1$, $k = 1, \ldots, m$. But

$$F^0(x) = \sum_{k=1}^{m} \alpha_k I_{x_k}(x)$$

is a distribution function which generates r^0 because

$$\int_0^1 r_i(x)\, dF^0(x) = \sum_{k=1}^{m} \alpha_k \int_0^1 r_i(x)\, dI_{x_k}(x)$$

$$= \sum_{k=1}^{m} \alpha_k r_i(x_k) = r_i^0$$

for $i = 1, \ldots, m$. Hence $r^0 \in R$.

By (1) and (2), $R = H$. Similarly, $S = H'$. $\qquad\qquad\qquad\qquad \square$

According to this theorem, in a separable game, for players 1 and 2 to choose mixed strategies amounts to their choosing points r and s in the convex hulls H and H', respectively. The expected payoff is again evaluated by (2.64).

Let $r^* = (r_1^*, \ldots, r_m^*) \in R$ and $s^* = (s_1^*, \ldots, s_n^*) \in S$ be respectively optimal strategies of players 1 and 2 in a separable game. The expected payoff to player 1 is

$$E(r^*, s^*) = \sum_{i=1}^{m} \sum_{j=1}^{n} a_{ij} r_i^* s_j^*. \qquad (2.70)$$

We have

$$\max_{r \in R} E(r, s^*) = E(r^*, s^*) = v = \min_{s \in S} E(r^*, s), \qquad (2.71)$$

where v is the value of the game.

How can the optimal strategies r^* and s^* be found? In what follows we shall describe a method for solution, called the *fixed point method* or *mapping method*.

Consider the following mapping. For any $r^* \in R$ (not necessarily optimal) we define the image of r^* in S under the mapping as

$$S(r^*) = \left\{ s: r^* \in R, \; E(r^*, s) = \min_{s \in S} E(r^*, s) \right\} \subseteq S.$$

Similarly, for any $s^* \in S$, define the image of s^* in R under the mapping as

$$R(s^*) = \left\{ r: s^* \in S, \; \max_{r \in R} E(r, s^*) = E(r, s^*) \right\} \subseteq R.$$

$S(r^*)$ and $R(s^*)$ are convex sets.

If $r^* \in R(s^*)$ and $s^* \in S(r^*)$, then by definition,

$$\max_{r \in R} E(r, s^*) = E(r^*, s^*), \qquad E(r^*, s^*) = \min_{s \in S} E(r^*, s).$$

That is, r^*, s^* satisfy (2.71). Hence they are optimal strategies of players 1 and 2, respectively.

Let R^* and S^* be the sets of all optimal strategies for players 1 and 2, respectively. If $r^* \in R^*$, we have

$$S(r^*) = \left\{ s: r^* \in R^*, \; E(r^*, s) = \min_{s \in S} E(r^*, s) \right\} = S^*.$$

That is to say, the image set of r^* is S^*. In other words, every $s^* \in S^*$ is an image of every $r^* \in R^*$. Analogously, every $r^* \in R^*$ is an image of every $s^* \in S^*$. Therefore, the optimal strategies $r^* \in R^*$ and $s^* \in S^*$ are all fixed points of the above mapping, and the problem of finding optimal strategies reduces to the problem of seeking for fixed points. We illustrate the method by an example.

Example 2.3. Suppose the payoff function of a separable game on the unit square $0 \leqslant x \leqslant 1$, $0 \leqslant y \leqslant 1$ is

$$P(x, y) = \left(\cos \frac{\pi}{2} x + \sin \frac{\pi}{2} x - 1 \right) y^2 + \frac{4}{3} \left(\cos \frac{\pi}{2} x - 3 \sin \frac{\pi}{2} x \right) y$$

$$+ \frac{1}{3} \left(5 \sin \frac{\pi}{2} x - 3 \cos \frac{\pi}{2} x \right).$$

The parametric equations of the curve C in (2.65) are

$$r_1 = \sin \frac{\pi}{2} x,$$
$$\qquad\qquad 0 \leqslant x \leqslant 1.$$
$$r_2 = \cos \frac{\pi}{2} x,$$

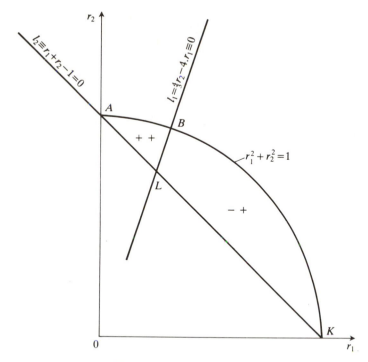

Figure 2.3. The set $R = H$

This is a circular arc, the equation being $r_1^2 + r_2^2 = 1$. The convex hull H of the curve C, i.e. the set R, is the region $ABKA$ shown in Fig. 2.3.

The parametric equations of the curve C' are

$$s_1 = y, \qquad 0 \leqslant y \leqslant 1.$$
$$s_2 = y^2,$$

This is a parabolic arc, the equation being $s_2 = s_1^2$. The convex hull H' of the curve C', i.e. the set S, is the region $0MN0$ in Fig. 2.4.

We write the payoff function (2.64) in the following two forms:

$$E(r, s) = \tfrac{4}{3}(r_2 - 3r_1)s_1 + (r_2 + r_1 - 1)s_2 + \tfrac{1}{3}(5r_1 - 3r_2) \qquad (2.72)$$
$$= (s_2 - 4s_1 + \tfrac{5}{3})r_1 + (\tfrac{4}{3}s_1 + s_2 - 1)r_2 - s_2. \qquad (2.73)$$

In Fig. 2.3, the straight lines $l_1 \equiv \tfrac{4}{3}r_2 - 4r_1 = 0$ and $l_2 \equiv r_1 + r_2 - 1 = 0$ divide player 1's strategy space $R = H$ into three regions:

ABL: $l_1 \geqslant 0, l_2 \geqslant 0$, excluding $l_1 = l_2 = 0$;
BKL: $l_1 \leqslant 0, l_2 \geqslant 0$, excluding $l_1 = l_2 = 0$;
   L: $l_1 = l_2 = 0$.

In Fig. 2.4, the straight lines $m_1 \equiv s_2 - 4s_1 + \tfrac{5}{3} = 0$ and $m_2 \equiv \tfrac{4}{3}s_1 + s_2 - 1 =$

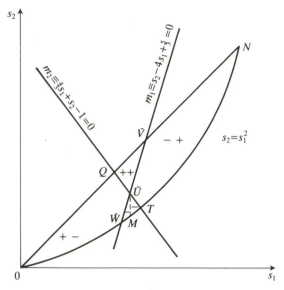

Figure 2.4. The set $S = H'$

0 divide player 2's strategy space $S = H'$ into five regions:

$$\begin{aligned}
QVU\colon & \quad m_1 \geqslant 0, \ m_2 \geqslant 0, \text{ excluding } m_1 = m_2 = 0; \\
0QUW\colon & \quad m_1 \geqslant 0, \ m_2 \leqslant 0, \text{ excluding } m_1 = m_2 = 0; \\
UVNT\colon & \quad m_1 \leqslant 0, \ m_2 \geqslant 0, \text{ excluding } m_1 = m_2 = 0; \\
WUT\colon & \quad m_1 \leqslant 0, \ m_2 \leqslant 0, \text{ excluding } m_1 = m_2; \\
UM\colon & \quad m_1 = m_2 \leqslant 0.
\end{aligned}$$

To find the fixed points, we examine each of the above regions and find its image under the mapping.

For example, let $r^0 \in ABL$. Since $l_1 \geqslant 0$, $l_2 \geqslant 0$ in ABL, $E(r^0, s)$ assumes its minimum at $s_1 = s_2 = 0$ or $s^0 = (0, 0)$. However, for the point $s^0 = 0 = (0, 0) \in 0QUW$, since $m_1 \geqslant 0$, $m_2 \leqslant 0$ in $0QUW$, we see that $E(r, s^0)$ assumes its maximum at $r_1 = 1$, $r_2 = 0$ or $r^0 = (1, 0) = K$. But the point K is outside ABL. Therefore, no point in ABL is a fixed point, i.e. there is no point in ABL which is an optimal strategy for player 1.

Let $r^0 \in BKL$. Then $l_1 \leqslant 0$, $l_2 \geqslant 0$. It is easily seen from (2.72) that, for a fixed r^0, if we keep s_1 unchanged, then $E(r^0, s)$ decreases as s_2 decreases; if we keep s_2 unchanged, then $E(r^0, s)$ decreases as s_1 increases. Therefore, $E(r^0, s)$ assumes its minimum on the arc $0TN$.

Let $s^0 \in 0QUW$. Since $m_1 \geqslant 0$, $m_2 \leqslant 0$ in $0QUW$, $E(r, s^0)$ assumes its maximum at $r_1 = 1$, $r_2 = 0$ or $r^0 = (1, 0)$. However, for the point $r^0 =$

$(1, 0) = K \in BKL$, since $l_1 \le 0$, $l_2 \ge 0$ in BKL, $E(r^0, s) = -4s_1 + \frac{5}{3}$ assumes its minimum at $s_1 = 1$, $s_2 = 1$ or $s^0 = (1, 1) = N$. But the point N is outside the region $0QUW$. Therefore, no point in $0QUW$ is an optimal strategy point of player 2.

Now let $s^0 \in UM$. Since $m_1 = m_2 \le 0$ on the segment UM, $E(r, s^0) = m_1(r_1 + r_2) - s_2$ assumes its maximum when $r_1 + r_2 = 1$. That is to say, every point on the line segment AK in R is an image point of s^0 under the mapping.

Again, let $s^0 \in WUT$. Then $m_1 \le 0$, $m_2 \le 0$, $m_1 \ne m_2$. It can be seen from (2.73) that $E(r, s^0)$ assumes its maximum on the line segment AK in R. The equation of AK is $r_1 + r_2 = 1$. Since $m_1 \ne m_2$, if $m_1 < m_2 \le 0$, $E(r, s^0)$ assumes its maximum at $r_2 = 1$ or the point A; if $m_2 < m_1 \le 0$, $E(r, s^0)$ assumes its maximum at $r_1 = 1$ or the point K. Therefore, the image point of $s^0 \in WUT$ under the mapping is either the point A or the point K of R.

Similarly, the image of every point of the other regions can also be found. Letting the symbol '$\to$' stand for 'map into,' we now list the complete mapping of the various regions of the two spaces as follows.

IN THE SPACE $R = H$:

Every point of $ABL \to 0$ of space S.
Each point of $BKL \to$ some point on the arc $0TN$.
$L \to$ every point of S.

IN THE SPACE $S = H'$:

Each point of $QVU \to$ some point on the arc ABK of R.
Every point of $0QUW \to$ point K of R.
Every point of $UVNT \to$ point A of R.
Each point of $WUT \to A$ or K.
Every point of $UM \to$ every point on the segment AK of R.

We now combine the above two mappings to obtain:

Each point of $QVU \to$ some point on the circular arc $ABK \to 0$ or some point on the arc $0TN$.
Every point of $0QUW \to K \to N$.
Every point of $UVNT \to A \to 0$.
Each point of $WUT \to A$ or $K \to 0$ or N.
Every point of $UM \to$ every point on the line segment AK, where

$$AL \text{ (excluding } L) \to 0,$$
$$LK \text{ (excluding } L) \to N,$$
$$L \to \text{every point of } S.$$

From the above analysis we see that only the following mappings

$$UM \rightarrow L \rightarrow UM$$

give rise to fixed points. Hence the solutions of the game are:

For player 1: L: $r_1^* = \frac{1}{4}$, $r_2^* = \frac{3}{4}$;
For player 2: UM: $s_1^* = \frac{1}{2}$, $\frac{1}{4} \leqslant s_2^* \leqslant \frac{1}{3}$.

The value of the game is $v = E(r^*, s^*) = -\frac{1}{3}$.

We now express the optimal strategies of the two players in terms of distribution functions. For player 1, when $x = 0$, $(r_1, r_2) = (0, 1)$. It follows from

$$r_1 = \int_0^1 \sin \frac{\pi}{2} x \, dI_0(x) = 0$$

and

$$r_2 = \int_0^1 \cos \frac{\pi}{2} x \, dI_0(x) = 1$$

that the point $(r_1, r_2) = (0, 1)$ is generated by the distribution function $I_0(x)$. Similarly, when $x = 1$, we have the point $(r_1, r_2) = (1, 0)$, which is generated by the distribution function $I_1(x)$. Therefore, the point

$$L = (r_1^*, r_2^*) = (\tfrac{1}{4}, \tfrac{3}{4}) = \tfrac{1}{4}(1, 0) + \tfrac{3}{4}(0, 1)$$

is generated by the distribution function

$$F^*(x) = \tfrac{1}{4}I_1(x) + \tfrac{3}{4}I_0(x). \tag{2.74}$$

This is the optimal strategy for player 1.

For player 2, when $y = 0$, the point $(s_1, s_2) = (0, 0)$ corresponds to the distribution function $I_0(y)$. When $y = t$, the point $(s_1, s_2) = (t, t^2)$ corresponds to the distribution function $I_t(y)$. For the points

$$s_1^* = \tfrac{1}{2}, \qquad \tfrac{1}{4} \leqslant s_2^* \leqslant \tfrac{1}{3},$$

since

$$(s_1, s_2) = \frac{1}{2t}(t, t^2) = (\tfrac{1}{2}, t/2)$$

is generated by the distribution function $(1/2t)I_t(y)$, we let

$$\tfrac{1}{4} \leqslant t/2 \leqslant \tfrac{1}{3},$$

so that

$$\tfrac{1}{2} \leqslant t \leqslant \tfrac{2}{3}.$$

Therefore, the distribution function expression of player 2's optimal strategy is

$$G^*(y) = \left(1 - \frac{1}{2t}\right)I_0(y) + \frac{1}{2t}I_t(y), \qquad \tfrac{1}{2} \leqslant t \leqslant \tfrac{2}{3}. \tag{2.75}$$

Finally, let us summarize the procedure for solving a separable game by means of the fixed point method or mapping method.
(1) Determine the curves C, C' and their convex hulls $H = R$, $H' = S$.
(2) Use the planes $l_j = 0$ to divide the space $R = H$ into regions

$$R_1, \ldots, R_i, \ldots, R_p.$$

Use the planes $m_i = 0$ to divide the space $S = H'$ into regions

$$S_1, \ldots, S_j, \ldots, S_q.$$

For every $r^0 \in R_i$, find the set of points $S(r^0) \subseteq S$ where the value $\min_{s \in S} E(r^0, s)$ is assumed. For every $s^0 \in S_j$, find the set of points $R(s^0) \subseteq R$ where the value $\max_{r \in R} E(r, s^0)$ is assumed.
(3) Find the fixed points

$$r^* \to s^* \to r^*.$$

These fixed points are the solution of the game. The value v of the game is also determined.
(4) Express the solution in terms of distribution functions.

2.7   An example of a game of timing

Consider an infinite game on the unit square $0 \leqslant x \leqslant 1$, $0 \leqslant y \leqslant 1$ with the payoff function

$$P(x, y) = \begin{cases} M_1(x, y), & \text{for } x > y, \\ M_0(x, y), & \text{for } x = y, \\ M_2(x, y), & \text{for } x < y, \end{cases} \tag{2.76}$$

where the function $P(x, y)$ is discontinuous along $x = y$. This type of game is called a *game of timing*.

During the period of the early development of game theory in the 1950s, investigations were made into many simplified military problems using the theory and technique of infinite games, especially those of games of timing, and interesting results were obtained.

In this section, only a simple example of a game of timimg will be discussed. It is customary to consider a problem of the classical duel as a model of this kind of game. Suppose each of players 1 and 2 has one bullet. If one player fires his bullet and misses, the other player knows at once that his opponent has no bullet left and can walk on until they are face to face and thus get a sure hit.

Suppose that players 1 and 2, starting at a distance one unit apart, walk toward each other. Each player may fire as he wishes when they are any distance apart. If a player hits his opponent, the 'payment' to the winner is 1, the loser has the payment -1. If both players fire at the same moment and both are hit or both survive, the payoff is 0.

A strategy for player 1 is to fire when the two players are at a distance x unit apart, $0 \leq x \leq 1$. Similarly, a strategy for player 2 is to fire when they are at a distance y unit apart, $0 \leq y \leq 1$. Let the accuracy function of player 1 be $P_1(x)$, $0 \leq x \leq 1$, which represents the probability of player 1's hitting his opponent if he fires when they are at a distance x unit apart. Normally the accuracy function is a decreasing function of the distance. Similarly, let $P_2(y)$, $0 \leq y \leq 1$, be the accuracy function of player 2. Both players wish to choose an appropriate opportunity to fire their bullets. If a player fires too early, the shot may not be accurate enough. On the other hand, if he holds his fire too long, his opponent may fire and hit him. The timing of the actions is decisive for both players. Hence the name.

We can regard the above duel as a zero-sum infinite game. The payoff to player 1 is

$$P(x, y) = \begin{cases} 1 \cdot P_1(x) + (-1)[1 - P_1(x)] = 2P_1(x) - 1, & x > y, \\ 1 \cdot P_1(x)[1 - P_2(x)] + (-1)[1 - P_1(x)]P_2(x) \\ \quad = P_1(x) - P_2(x), & x = y, \\ (-1)P_2(y) + 1 \cdot [1 - P_2(y)] = 1 - 2P_2(y), & x < y. \end{cases}$$

In this formulation, player 1 fires when he is x unit distant from player 2. $x > y$ means player 1 fires before player 2 fires. $P_1(x)$ is the probability of his hitting player 2; if player 2 is hit, the payoff to player 1 is 1. $1 - P_1(x)$ is the probability that player 1 fires and misses; if player 2 is not hit, then player 1 is sure to be hit, and the payoff to player 1 is -1. Hence, the sum of the two terms $1 \cdot P_1(x)$ and $(-1)[1 - P_1(x)]$ is the expected payoff to player 1. In the case $x = y$ the two players fire at the same time. The term $1 \cdot P_1(x)[1 - P_2(x)]$ is the expected payoff to player 1 if player 2 is hit but player 1 is not hit. The other term $(-1)[1 - P_1(x)]P_2(x)$ is the expected payoff if player 2 is not hit but player 1 is hit. Finally $x < y$ is the case in which player 2 fires before player 1 fires.

Let us evaluate

$$\max_{0 \leq x \leq 1} \min_{0 \leq y \leq 1} P(x, y).$$

Since P_1, P_2 are decreasing functions of their respective arguments, we have $1 - 2P_2(y) > 1 - 2P_2(x)$ for $x < y$. Hence

$$\max_{0 \leqslant x \leqslant 1} \min_{0 \leqslant y \leqslant 1} P(x, y)$$

$$= \max_{0 \leqslant x \leqslant 1} \min\{2P_1(x) - 1, \ P_1(x) - P_2(x), \ 1 - 2P_2(x)\}.$$

Divide $0 \leqslant x \leqslant 1$ into three intervals as follows:

$$A = \{x \colon P_1(x) + P_2(x) \geqslant 1\},$$
$$B = \{x \colon P_1(x) + P_2(x) = 1\},$$
$$C = \{x \colon P_1(x) + P_2(x) \leqslant 1\}.$$

Let

$$\mu(x) = \min\{2P_1(x) - 1, \ P_1(x) - P_2(x), \ 1 - 2P_2(x)\}.$$

Then

$$\max_{0 \leqslant x \leqslant 1} \min_{0 \leqslant y \leqslant 1} P(x, y) = \max_{0 \leqslant x \leqslant 1} \mu(x)$$

$$= \max\left\{\max_{x \in A} \mu(x), \ \max_{x \in B} \mu(x), \ \max_{x \in C} \mu(x)\right\}.$$

Let x^* be such that $P_1(x^*) + P_2(x^*) = 1$, as shown in Fig. 2.5. Then

$$A = [0, x^*], \qquad B = \{x^*\}, \qquad C = [x^*, 1].$$

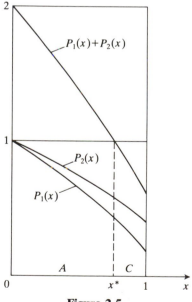

Figure 2.5

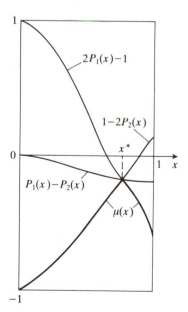

Figure 2.6

(1) When $x \in A$,
$$P_1(x) + P_2(x) \geq 1.$$

We have
$$1 - 2P_2(x) \leq P_1(x) - P_2(x) \leq 2P_1(x) - 1.$$

Hence
$$\mu(x) = 1 - 2P_2(x),$$

which is an increasing function of x, see Fig. 2.6. Therefore,
$$\max_{x \in A} \mu(x) = 1 - 2P_2(x^*).$$

(2) When $x \in B$,
$$P_1(x) + P_2(x) = 1.$$

We have
$$1 - 2P_2(x) = P_1(x) - P_2(x) = 2P_1(x) - 1.$$

Hence
$$\mu(x) = P_1(x) - P_2(x),$$

and
$$\max_{x \in B} \mu(x) = P_1(x^*) - P_2(x^*).$$

(3) When $x \in C$,
$$P_1(x) + P_2(x) \leq 1.$$

We have
$$2P_1(x) - 1 \leq P_1(x) - P_2(x) \leq 1 - 2P_2(x).$$

Hence
$$\mu(x) = 2P_1(x) - 1,$$

which is an decreasing function of x. Therefore,
$$\max_{x \in C} \mu(x) = 2P_1(x^*) - 1.$$

It follows from (1), (2), (3) that
$$\max_{x \in A} \mu(x) = \max_{x \in B} \mu(x) = \max_{x \in C} \mu(x) = P_1(x^*) - P_2(x^*).$$

Hence
$$\max_{0 \leq x \leq 1} \min_{0 \leq y \leq 1} P(x, y) = P_1(x^*) - P_2(x^*).$$

In a similar manner it can be proved that

$$\min_{0 \leqslant y \leqslant 1} \max_{0 \leqslant x \leqslant 1} P(x, y) = P_1(y^*) - P_2(y^*),$$

where y^* satisfies the equation

$$P_1(y^*) + P_2(y^*) = 1.$$

Thus we see that the game has a saddle point (x^*, y^*), $x^* = y^*$, satisfying

$$P_1(x^*) + P_2(x^*) = 1.$$

The two players should fire simultaneously when they are at a distance x^* unit apart. The value of the game is

$$v = P_1(x^*) - P_2(x^*).$$

Example 2.4. Assume that the accuracy function of player 1 is $P_1(x) = 1 - x^2$, the accuracy function of player 2 is $P_2(y) = 1 - y$. Then the payoff function of the duel game is

$$P(x, y) = \begin{cases} 1 - 2x^2, & x > y, \\ x - x^2, & x = y, \\ 2y - 1, & x < y. \end{cases}$$

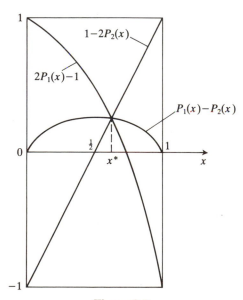

Figure 2.7

According to the above analysis, the optimal strategies for the two duelists are to fire at the same moment when they are at a distance x^* unit apart, where x^* satisfies the equation

$$P_1(x^*) + P_2(x^*) = 1,$$

or

$$1 - x^{*2} + 1 - x^* = 1,$$

or

$$x^{*2} + x^* = 1.$$

We obtain

$$x^* = 0.618.$$

Hence the value of the game is

$$v = P_1(x^*) - P_2(x^*) = x^* - x^{*2} = 0.236;$$

see Fig. 2.7.

3

N-PERSON NON-COOPERATIVE GAMES

3.1 Introduction

All the games discussed in the preceding chapters are games with two players. Moreover, all those games are zero-sum games. The interests of the two players in a zero-sum two-person game are necessarily conflicting: a situation or an outcome which is advantageous to one player is certainly disadvantageous to the other. Each player seeks a strategy which is most advantageous to himself; this strategy is at the same time a strategy which causes the greatest loss to the other player.

We shall now turn to the discussion of *n*-person games. *N*-person games (for $n \geq 2$) will be classified into two classes: non-cooperative games and cooperative games. This chapter deals with non-cooperative games.

In a non-cooperative game, cooperation between players is forbidden. That is to say, no pre-play communications such as binding agreements, correlation of strategies, or side payments are permitted. Every player seeks to maximize his own individual payoff and is in search for a strategy which is to him the most advantageous. However, in an *n*-person non-cooperative game, what is advantageous to one player may not be disadvantageous to some of the other players. Even in a two-person non-cooperative game, the interests of the two players may not be completely conflicting. Of course, in this case the game is no longer zero-sum, because zero-sum two-person games must necessarily be conflicting.

We begin by examining two well-known examples of the two-person non-cooperative game in the literature of game theory.

Example 3.1. (The prisoners' dilemma). Two prisoners are suspected of taking part in a serious crime and shut up in separate cells. The punishment depends on whether they confess or not. Each prisoner has two strategies: to confess (strategy 2) or not to confess (strategy 1). If both confess, they will be sentenced for six years. If neither confesses, both will get a sentence for one year on account of a lesser guilt. If one confesses and the other does not, the former will be confined only for three months, while the latter will receive a severe sentence of ten years. Since communications between the prisoners are strictly forbidden, no cooperation is possible.

The outcomes can be listed in the table below.

                                    Prisoner 2

                          not confess   confess

            not confess ⎡ 1: 1 year   1: 10 years ⎤
                        ⎢ 2: 1 year   2: 3 months ⎥
Prisoner 1   ⎢   ⎥
            confess   ⎢ 1: 3 months   1: 6 years   ⎥
                        ⎣ 2: 10 years   2: 6 years   ⎦

If every punishment is estimated by a number, the game can be represented by the following table.

                            Player 2
                          1   2

Player 1   1 ⎡ (8, 8)   (0, 10) ⎤
            2 ⎣ (10, 0) (2, 2) ⎦ .

Here, for each player strategy 1 represents 'not confess' and strategy 2 represents 'confess'. For every pair of numbers in the matrix (e.g. (0, 10)), the first number (0) is the payoff to player 1, the second number (10) is the payoff to player 2. (0, 10) are the outcomes to the two players when player 1 chooses his strategy 1 and player 2 chooses his strategy 2.

It is evident that this game is not a zero-sum game.

We can also write the outcomes into two payoff matrices, one for each player:

$$A = \begin{bmatrix} 8 & 0 \\ 10 & 2 \end{bmatrix}, \qquad B = \begin{bmatrix} 8 & 10 \\ 0 & 2 \end{bmatrix}.$$

A is player 1's payoff matrix, B is player 2's payoff matrix. For example, if player 1 chooses his strategy 2 and player 2 chooses his strategy 1, then the payoffs to players 1 and 2 are respectively $a_{21} = 10$ and $b_{21} = 0$.

The outcomes of this example can also be estimated by the elements of the following payoff matrices:

$$A = \begin{bmatrix} -12 & -120 \\ -3 & -72 \end{bmatrix}, \qquad B = \begin{bmatrix} -12 & -3 \\ -120 & -72 \end{bmatrix}.$$

Example 3.2 (The battle of the sexes). On Saturday evening the husband (player 1) wishes to go to the football match and the wife (player 2) to the ballet. Neither likes to go alone to his or her own preferred choice. However, the rules of this non-cooperative game forbid any pre-play communication and they must make their choices independently.

Let strategy 1 denote the choice of football match and strategy 2 that of ballet for both players. The satisfaction derived in each case may be estimated by the elements in the following payoff matrices:

$$A = \begin{bmatrix} 2 & -1 \\ -1 & 1 \end{bmatrix}, \quad B = \begin{bmatrix} 1 & -1 \\ -1 & 2 \end{bmatrix}.$$

In each of the above examples, there are two players. They choose their strategies without any pre-play communication, without any side payments. This kind of game is called a two-person *non-cooperative game*.

If in a two-person non-cooperative game each player has more than two strategies, for example, player 1 has m strategies, player 2 has n strategies, then the payoff matrices are

$$A = \begin{bmatrix} a_{11} & \cdots & a_{1n} \\ \vdots & & \vdots \\ a_{m1} & \cdots & a_{mn} \end{bmatrix}, \quad B = \begin{bmatrix} b_{11} & \cdots & b_{1n} \\ \vdots & & \vdots \\ b_{m1} & \cdots & b_{mn} \end{bmatrix}.$$

This kind of game is called a *bimatrix game*. The games in the two examples above are the simplest of the bimatrix games.

When the number of players in a non-cooperative game is n, the game is called an *n-person non-cooperative game*. An n-person non-cooperative game is characterized by the following elements:

(1) The set of players $I = \{1, \ldots, n\}$.
(2) Every player i has a finite set of pure strategies:

$$S_i = \{s^{(i)}\} = \{s_1^{(i)}, \ldots, s_{m_i}^{(i)}\}, \quad i = 1, \ldots, n.$$

(3) Every player i has a payoff function P_i, $i = 1, \ldots, n$.

If each player i chooses a strategy $s^{(i)}$, the n-tuple

$$s = (s^{(1)}, \ldots, s^{(n)}),$$

where $s^{(i)} \in S_i$, $i = 1, \ldots, n$, is called a *situation* of the game. For each situation $s = (s^{(1)}, \ldots, s^{(n)})$ of the game, player i gets the payoff

$$P_i = P_i(s), \quad i = 1, \ldots, n.$$

These are the payoff functions of the game in pure strategies.

Thus an n-person non-cooperative game Γ can be represented by the following symbol:

$$\Gamma \equiv [I, \{S_i\}, \{P_i\}], \tag{3.1}$$

where $I = \{1, \ldots, n\}$, $\{S_i\} = \{S_1, \ldots, S_n\}$, and $\{P_i\} = \{P_1, \ldots, P_n\}$.

We now introduce the following notation:

$$s \parallel t^{(i)} = (s^{(1)}, \ldots, s^{(i-1)}, t^{(i)}, s^{(i+1)}, \ldots, s^{(n)}). \qquad (3.2)$$

It is defined as the situation obtained from the situation $s = (s^{(1)}, \ldots, s^{(n)})$ in which the strategy $s^{(i)}$ of player i is replaced by the strategy $t^{(i)}$, the other players' strategies remaining unchanged. Obviously we have $s \parallel s^{(i)} = s$.

Definition 3.1. Let s^* be a situation of an n-person non-cooperative game (3.1). If for every $i \in I$ and every $s^{(i)} \in S_i$ ($s^{(i)} = s_k^{(i)}$, $k = 1, \ldots, m_i$) we have

$$P_i(s^* \parallel s^{(i)}) \leqslant P_i(s^*), \qquad (3.3)$$

then s^* is called an *equilibrium situation* or *equilibrium point* of Γ.

In Example 3.1, the situation $s^* = (2, 2)$, consisting of player 1's strategy 2 and player 2's strategy 2, is an equilibrium situation:

$$0 = a_{12} \leqslant a_{22} = 2, \qquad 0 = b_{21} \leqslant b_{22} = 2.$$

It is clear that in an n-person non-cooperative game, equilibrium point may not exist. Just as in the case of matrix games, it is necessary to consider mixed strategies for the players in a non-cooperative game.

For every $i \in I$, let $x^{(i)}$ be a mixed strategy of player i, i.e. a probability distribution defined on S_i. That is

$$x^{(i)} = (x_1^{(i)}, \ldots, x_{m_i}^{(i)}),$$

where

$$x_k^{(i)} \geqslant 0, \qquad k = 1, \ldots, m_i, \qquad \sum_{k=1}^{m_i} x_k^{(i)} = 1. \qquad (3.4)$$

Player i chooses his pure strategy $s_k^{(i)}$ with probability $x_k^{(i)}$, $k = 1, \ldots, m_i$. We call $x = (x^{(1)}, \ldots, x^{(n)})$ a *mixed strategy situation* of Γ.

Analogously to (3.2), we define

$$x \parallel z^{(i)} = (x^{(1)}, \ldots, x^{(i-1)}, z^{(i)}, x^{(i+1)}, \ldots, x^{(n)}). \qquad (3.5)$$

This means that in the mixed strategy situation

$$x = (x^{(1)}, \ldots, x^{(n)})$$

player i has replaced his mixed strategy $x^{(i)}$ by another mixed strategy $z^{(i)}$, the strategies of the other players remaining unchanged. The resultant mixed strategy situation is $x \parallel z^{(i)}$. Obviously we have $x \parallel x^{(i)} = x$.

An n-person non-cooperative game in mixed strategies is characterized by the following elements:

(1) The set of players $I = \{1, \ldots, n\}$.

(2) Every player i has a set of mixed strategies:

$$X_i = \{x^{(i)}\} = \{(x_1^{(i)}, \ldots, x_{m_i}^{(i)})\}, \qquad i = 1, \ldots, n,$$

where $x_1^{(i)}, \ldots, x_{m_i}^{(i)}$ satisfy (3.4).

(3) Every player i has a payoff function $P_i = P_i(s)$, $i = 1, \ldots, n$. The expected payoff for player i is denoted by $E_i = E_i(x)$, where $x = (x^{(1)}, \ldots, x^{(n)})$ is the situation in mixed strategies.

Thus an n-person non-cooperative game Γ in mixed strategies can be represented by

$$\Gamma \equiv [I, \{X_i\}, \{P_i\}], \tag{3.6}$$

where

$$I = \{1, \ldots, n\}; \qquad \{X_i\} = \{X_1, \ldots, X_n\};$$
$$X_i = \{x^{(i)}\} = \{(x_1^{(i)}, \ldots, x_{m_i}^{(i)})\}; \qquad \{P_i\} = \{P_1, \ldots, P_n\}.$$

For simplicity we use the same letter Γ to denote a game in mixed strategies. We shall denote the expected payoffs by $\{E_i\}$.

In passing, we point out that the X_i here is the $(m_i - 1)$-dimensional simplex S_{m_i} in the m_i-dimensional Euclidean space in Chapter 1.

Definition 3.2. Let x^* be a mixed strategy situation of an n-person non-cooperative game (3.6). If for every $i \in I$ and every $x^{(i)} \in X_i$ we have

$$E_i(x^* \| x^{(i)}) \leqslant E_i(*), \tag{3.7}$$

then x^* is called an *equilibrium situation* or *equilibrium point* of Γ (in mixed strategies).

J. F. Nash proved that equilibrium point in mixed strategies always exists. (See Nash, 1950, 1951; Luce and Raiffa, 1957; Vorob'ev, 1977.)

3.2   Existence of equilibrium point: Nash's theorem

Let us first prove the following important property of an equilibrium point.

Theorem 3.1. *Let* $\Gamma \equiv [I, \{X_i\}, \{P_i\}]$ *be an* n-*person non-cooperative game. A necessary and sufficient condition for* x^* *to be an equilibrium point of* Γ *is that for every player* i *and every pure strategy* $s^{(i)} \in S_i$ *we have*

$$E_i(x^* \| s^{(i)}) \leqslant E_i(x^*). \tag{3.8}$$

Here $E_i(x^* \| s^{(i)})$ *is the expected payoff when the mixed strategy* $x^{*(i)}$ *of player* i *is replaced by the pure strategy* $s^{(i)}$.

Proof. The necessity of the condition follows directly from the definition of an equilibrium point.

To prove the sufficiency of the condition, suppose that (3.8) holds, i.e. for every i we have

$$E_i(x^* \| s_k^{(i)}) \leqslant E_i(x^*), \qquad k = 1, \ldots, m_i. \tag{3.9}$$

Let $x^{(i)} = (x_1^{(i)}, \ldots, x_{m_i}^{(i)}) \in X_i$ be any mixed strategy of player i. Multiplying both sides of the m_i successive inequalities in (3.9), respectively, by the factors $x_k^{(i)}$, $k = 1, \ldots, m_i$, we obtain

$$E_i(x^* \| s_k^{(i)}) x_k^{(i)} \leqslant E_i(x^*) x_k^{(i)}, \qquad k = 1, \ldots, m_i.$$

Summing for $k = 1, \ldots, m_i$:

$$\sum_{k=1}^{m_i} E_i(x^* \| s_k^{(i)}) x_k^{(i)} \leqslant E_i(x^*) \sum_{k=1}^{m_i} x_k^{(i)}.$$

The left-hand side of this inequality is $E_i(x^* \| x^{(i)})$. The sum in the right-hand side is equal to 1. Hence

$$E_i(x^* \| x^{(i)}) \leqslant E_i(x^*), \qquad i = 1, \ldots, n.$$

This proves that x^* is an equilibrium point of Γ. □

Using this theorem we can examine whether a mixed strategy x is an equilibrium point. For example, in Example 3.2 the payoffs to the two players are

$$\begin{bmatrix} (2, 1) & (-1, -1) \\ (-1, -1) & (1, 2) \end{bmatrix}.$$

It is easily verified that

$$x^{(1)} = (\tfrac{3}{5}, \tfrac{2}{5})$$

and

$$x^{(2)} = (\tfrac{2}{5}, \tfrac{3}{5})$$

form an equilibrium point

$$x^* = ((\tfrac{3}{5}, \tfrac{2}{5}), (\tfrac{2}{5}, \tfrac{3}{5})).$$

In fact, we have $E_1(x^*) = \tfrac{1}{5}$, $E_2(x^*) = \tfrac{1}{5}$;

$$E_1(x^* \| s_1^{(1)}) = \tfrac{1}{5} \leqslant E_1(x^*),$$
$$E_1(x^* \| s_2^{(1)}) = \tfrac{1}{5} \leqslant E_1(x^*),$$
$$E_2(x^* \| s_1^{(2)}) = \tfrac{1}{5} \leqslant E_2(x^*),$$
$$E_2(x^* \| s_2^{(2)}) = \tfrac{1}{5} \leqslant E_2(x^*).$$

Therefore, x^* is an equilibrium point of the game. See Example 3.3 below.

The following is Nash's fundamental theorem on the equilibrium points of n-person non-cooperative games.

Theorem 3.2. *Every n-person non-cooperative game*

$$\Gamma \equiv [I, \{X_i\}, \{P_i\}]$$

has an equilibrium point.

Proof. Let $x = (x^{(1)}, \ldots, x^{(n)})$ be an arbitrary mixed strategy situation of Γ. For every pure strategy $s_j^{(i)}$, $j = 1, \ldots, m_i$, of every player $i \in I = \{1, \ldots, n\}$, define

$$\phi_{ij}(x) = \max\{0, E_i(x \| s_j^{(i)}) - E_i(x)\}. \tag{3.10}$$

For every $x_j^{(i)}$, $j = 1, \ldots, m_i$, $i = 1, \ldots, n$, define

$$y_j^{(i)} = [x_j^{(i)} + \phi_{ij}(x)] \Big/ \Big[1 + \sum_{j=1}^{m_i} \phi_{ij}(x)\Big]. \tag{3.11}$$

The numerator of the right-hand side of (3.11) is ≥ 0, and the denominator is ≥ 1. Hence

$$y_j^{(i)} \geq 0, \qquad j = 1, \ldots, m_i, \qquad \sum_{j=1}^{m_i} y_j^{(i)} = 1,$$

and therefore $y^{(i)} = (y_1^{(i)}, \ldots, y_{m_i}^{(i)})$ is a mixed strategy of player i. $y^{(i)} = (y_1^{(i)}, \ldots, y_{m_i}^{(i)})$ is a continuous function of x, hence $y = (y^{(1)}, \ldots, y^{(n)})$ is a continuous function of x. By Brouwer's fixed point theorem (which states that a continuous function defined on a compact convex subset S of a finite-dimensional Euclidean space mapping S into itself has at least one fixed point), there exists a fixed point

$$x^* = (x^{*(1)}, \ldots, x^{*(n)}),$$

where

$$x^{*(i)} = (x_1^{*(i)}, \ldots, x_{m_i}^{*(i)}) \in X_i,$$

such that

$$x_j^{*(i)} = [x_j^{*(i)} + \phi_{ij}(x^*)] \Big/ \Big[1 + \sum_{j=1}^{m_i} \phi_{ij}(x^*)\Big], \qquad \begin{matrix} j = 1, \ldots, m_i, \\ i = 1, \ldots, n. \end{matrix} \tag{3.12}$$

We shall prove that this fixed point x^* is an equilibrium point of the game Γ.

In the first place, for the above arbitrary mixed strategy situation $x = (x^{(1)}, \ldots, x^{(n)})$, the mixed strategy $x^{(i)}$ of player i will use a certain pure strategy $s_k^{(i)}$ such that

$$E_i(x \parallel s_k^{(i)}) \leqslant E_i(x). \tag{3.13}$$

Proof. For every $x^{(i)} = (x_1^{(i)}, \ldots, x_{m_i}^{(i)})$, there exist some (at least one) j such that $x_j^{(i)} > 0$. That is to say, every mixed strategy $x^{(i)}$ must contain some pure strategies. Let $s_k^{(i)}$ be one such pure strategy which is 'least profitable' to player i, i.e.

$$E_i(x \parallel s_k^{(i)}) \leqslant E_i(x \parallel s_j^{(i)}),$$

where j satisfies the condition $x_j^{(i)} > 0$. Multiplying both sides of the inequalities respectively by x_j for all j, including those j for which $x_j^{(i)} > 0$ and those for which $x_j^{(i)} = 0$, and summing over $j = 1, \ldots, m_i$, we obtain

$$E_i(x \parallel s_k^{(i)}) \sum_{j=1}^{m_i} x_j^{(i)} \leqslant \sum_{\substack{j \\ x_j^{(i)} > 0}} E_i(x \parallel s_j^{(i)}) x_j^{(i)} + \sum_{\substack{j \\ x_j^{(i)} = 0}} E_i(x \parallel s_j^{(i)}) x_j^{(i)}$$

$$= \sum_{j=1}^{m_i} E_i(x \parallel s_j^{(i)}) x_j^{(i)} = E_i(x \parallel x^{(i)}) = E_i(x).$$

This proves that $E_i(x \parallel s_k^{(i)}) \leqslant E_i(x)$.

Hence, for $x^* = (x^{*(1)}, \ldots, x^{*(n)})$, player i's strategy $x^{*(i)}$ must have a component $x_k^{*(i)} > 0$ such that

$$E_i(x^* \parallel s_k^{(i)}) \leqslant E_i(x^*),$$

or

$$E_i(x^* \parallel s_k^{(i)}) - E_i(x^*) \leqslant 0.$$

By (3.10), $\phi_{ik}(x^*) = 0$. For the above strategy $x_k^{*(i)}$ of player i, (3.12) becomes

$$x_k^{*(i)} = [x_k^{*(i)} + \phi_{ik}(x^*)] \bigg/ \left[1 + \sum_{j=1}^{m_i} \phi_{ij}(x^*)\right]$$

$$= [x_k^{*(i)}] \bigg/ \left[1 + \sum_{j=1}^{m_i} \phi_{ij}(x^*)\right],$$

where $x_k^{*(i)} > 0$. It follows that

$$\sum_{j=1}^{m_i} \phi_{ij}(x^*) = 0.$$

By the definition (3.10) of $\phi_{ij}(x^*)$, all $\phi_{ij}(x^*)$ are non-negative; hence the above equation implies

$$\phi_{ij}(x^*) = 0$$

for all $j = 1, \ldots, m_i$. Using (3.10) once more, we have

$$0 \geqslant E_i(x^* \| s_j^{(i)}) - E_i(x^*), \qquad j = 1, \ldots, m_i,$$

i.e.

$$E_i(x^* \| s_j^{(i)}) \leqslant E_i(x^*), \qquad j = 1, \ldots, m_i.$$

These inequalities hold for all $i = 1, \ldots, n$. By Theorem 3.1, x^* is an equilibrium point of Γ. $\square$

3.3   Equilibrium points of 2×2 bimatrix games

The determination of equilibrium points of a bimatrix game with large m, n is quite difficult. Our discussion will be restricted to the case of 2×2 bimatrix games only.

Let the payoff matrices of players 1 and 2 in a bimatrix game be respectively

$$A = \begin{bmatrix} a & b \\ c & d \end{bmatrix}, \qquad B = \begin{bmatrix} a' & b' \\ c' & d' \end{bmatrix}.$$

Denote by $X = (x, 1 - x)$ and $Y = (y, 1 - y)$ mixed strategies of players 1 and 2 respectively, where $0 \leqslant x \leqslant 1, \ 0 \leqslant y \leqslant 1$.

A mixed strategy situation of the game is completely determined by the pair of numbers (x, y). For simplicity, the expected payoffs to players 1 and 2 corresponding to the mixed strategies (x, y) will be denoted by $E_1(x, y)$ and $E_2(x, y)$, respectively.

By Theorem 3.1 a necessary and sufficient condition for (x, y) to be an equilibrium point of the game is that

$$E_1(1, y) \leqslant E_1(x, y), \tag{3.14}$$
$$E_1(0, y) \leqslant E_1(x, y), \tag{3.15}$$
$$E_2(x, 1) \leqslant E_2(x, y), \tag{3.16}$$
$$E_2(x, 0) \leqslant E_2(x, y). \tag{3.17}$$

Let us solve the inequalities (3.14) and (3.15) first. Since

$$E_1(x, y) = XAY^t,$$

the two inequalities reduce to

$$Q(1 - x)y - q(1 - x) \leqslant 0, \tag{3.18}$$
$$Qxy - qx \geqslant 0, \tag{3.19}$$

where

$$Q = a - b - c + d, \qquad q = d - b. \tag{3.20}$$

If $Q = 0$ and $q = 0$, the solutions of (3.18), (3.19) are

$$\text{all} \quad x \in [0, 1] \quad \text{and} \quad \text{all} \quad y \in [0, 1]. \tag{3.21}$$

If $Q = 0$, $q > 0$, the solutions are

$$x = 0, \quad 0 \leqslant y \leqslant 1. \tag{3.22}$$

If $Q = 0$, $q < 0$, the solutions are

$$x = 1, \quad 0 \leqslant y \leqslant 1. \tag{3.23}$$

If $Q \neq 0$, the solutions are (it is assumed that $Q > 0$; the case $Q < 0$ is similar)

$$\begin{aligned} x &= 0, & y &\leqslant q/Q = \alpha, \\ 0 &< x < 1, & y &= q/Q = \alpha, \\ x &= 1, & y &\geqslant q/Q = \alpha. \end{aligned} \tag{3.24}$$

Similarly, let

$$R = a' - b' - c' + d', \qquad r = d' - c', \tag{3.25}$$

and denote

$$\frac{r}{R} = \beta.$$

Then the inequalities (3.16) and (3.17) reduce to

$$Rx(1 - y) - r(1 - y) \leqslant 0, \tag{3.26}$$

$$Rxy - ry \geqslant 0. \tag{3.27}$$

If $R = 0$ and $r = 0$, the solutions of (3.26), (3.27) are

$$\text{all} \quad x \in [0, 1] \quad \text{and} \quad \text{all} \quad y \in [0, 1]. \tag{3.28}$$

If $R = 0$, $r > 0$, the solutions are

$$0 \leqslant x \leqslant 1, \quad y = 0. \tag{3.29}$$

If $R = 0$, $r < 0$, the solutions are

$$0 \leqslant x \leqslant 1, \quad y = 1. \tag{3.30}$$

If $R \neq 0$, the solutions are (it is assumed that $R > 0$; the case $R < 0$ is similar)

$$\left. \begin{aligned} x &\leqslant r/R = \beta, & y &= 0, \\ x &= r/R = \beta, & 0 &< y < 1, \\ x &\geqslant r/R = \beta, & y &= 1. \end{aligned} \right\} \tag{3.31}$$

Combining (3.21)–(3.24) and (3.28)–(3.31), we obtain the equilibrium points of the 2 × 2 bimatrix game in all cases.

Example 3.3. For the 2 × 2 bimatrix game of Example 3.2,

$$A = \begin{bmatrix} 2 & -1 \\ -1 & 1 \end{bmatrix}, \qquad B = \begin{bmatrix} 1 & -1 \\ -1 & 2 \end{bmatrix}. \qquad (3.32)$$

Equations (3.20) and (3.25) give

$$Q = 5 > 0, \qquad q = 2, \qquad \alpha = q/Q = \tfrac{2}{5},$$
$$R = 5 > 0, \qquad r = 3, \qquad \beta = r/R = \tfrac{3}{5}.$$

Substituting these values into (3.24) and (3.31), we have

$$\left.\begin{array}{ll} x = 0, & y \leqslant \tfrac{2}{5}, \\ 0 < x < 1, & y = \tfrac{2}{5}, \\ x = 1, & y \geqslant \tfrac{2}{5}; \end{array}\right\} \qquad (3.33)$$

$$\left.\begin{array}{ll} x \leqslant \tfrac{3}{5}, & y = 0, \\ x = \tfrac{3}{5}, & 0 < y < 1, \\ x \geqslant \tfrac{3}{5}, & y = 1. \end{array}\right\} \qquad (3.34)$$

Solving these inequalities, we see that the game has three equilibrium points:

$$(x, y) = (0, 0), \quad (\tfrac{3}{5}, \tfrac{2}{5}), \quad (1, 1).$$

Expressed in terms of the players' mixed strategies

$$(X, Y) = ((x, 1-x), (y, 1-y)),$$

the three equilibrium points are

$$((0, 1), (0, 1)), \qquad (3.35)$$
$$((\tfrac{3}{5}, \tfrac{2}{5}), (\tfrac{2}{5}, \tfrac{3}{5})), \qquad (3.36)$$
$$((1, 0), (1, 0)). \qquad (3.37)$$

The first equilibrium point is the case where players 1 and 2 both choose their pure strategy 2. The second equilibrium point is the case where player 1 chooses strategy 1 with probability $\tfrac{3}{5}$ and strategy 2 with probability $\tfrac{2}{5}$, player 2 chooses strategy 1 with probability $\tfrac{2}{5}$ and strategy 2 with probability $\tfrac{3}{5}$. The third equilibrium point is the case where both players choose strategy 1.

The set of all situations of the inequalities (3.33) is represented in Fig. 3.1 by the heavy lines; that for (3.34) is indicated by the dashed lines. It is easily seen that the points of intersection of the heavy lines and the dashed lines are the three equilibrium points.

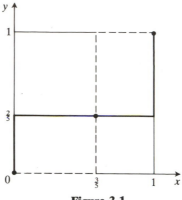

Figure 3.1

We have seen in Example 3.2 that the expected payoffs to players 1 and 2 at the second equilibrium point (3.36) are respectively

$$E_1 = \tfrac{1}{5}, \qquad E_2 = \tfrac{1}{5}.$$

These payoffs are obviously less than the payoffs to players 1 and 2 at the first and the third equilibrium points. However, since the game is a two-person non-cooperative game, any pre-play agreements between the players are forbidden, and they cannot be sure to arrive at either the first or the third equilibrium situation.

In this problem of 'battle of the sexes' (3.32), when players 1, 2 use respectively mixed strategies $(x, 1-x)$, $(y, 1-y)$, $0 \leqslant x \leqslant 1$, $0 \leqslant y \leqslant 1$, their expected payoffs are

$$E_1(x, y) = 5xy - 2(x + y) + 1, \tag{3.38}$$

$$E_2(x, y) = 5xy - 3(x + y) + 2. \tag{3.39}$$

In Fig. 3.2 we give the graph of the set of points (E_1, E_2) for $0 \leqslant x \leqslant 1$, $0 \leqslant y \leqslant 1$ in the $E_1 E_2$ plane. From (3.38), (3.39) we have

$$5(E_1 - E_2)^2 - 2(E_1 + E_2) + 1 = 5(x - y)^2.$$

when $0 \leqslant x \leqslant 1$, $0 \leqslant y \leqslant 1$, the following inequalities hold:

$$0 \leqslant 5(E_1 - E_2)^2 - 2(E_1 + E_2) + 1 \leqslant 5. \tag{3.40}$$

Furthermore, if we solve for x and y from (3.38) and (3.39), we must have

$$2E_1 - 3E_2 - 1 \leqslant 0, \tag{3.41}$$

$$3E_1 - 2E_2 + 1 \geqslant 0. \tag{3.42}$$

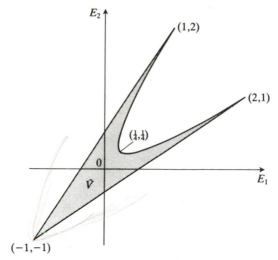

Figure 3.2

Consider (3.40). The graph of the equation

$$5(E_1 - E_2)^2 - 2(E_1 + E_2) + 1 = 0$$

is a parabola with the vertex at $(\frac{1}{4}, \frac{1}{4})$ and passing through the points $(1, 2)$, $(2, 1)$. The points (E_1, E_2) satisfying the inequality

$$5(E_1 - E_2)^2 - 2(E_1 + E_2) + 1 \geqslant 0$$

lie on or outside the parabola, i.e. in that side of the parabola which contains the origin 0.

Thus the points (E_1, E_2) satisfying (3.40), (3.41) and (3.42) form a region. It is the shaded region V in Fig. 3.2. To every pair of mixed strategies (x, y) of players 1 and 2, there corresponds a point (E_1, E_2) of V; on the other hand, to every point (E_1, E_2) of V, there corresponds at least one pair of mixed strategies (x, y) with this point (E_1, E_2) as the players' expected payoffs. The region V is called the *payoff set* of the game.

Although there are three equilibrium points for this game, none of them can be treated as a satisfactory solution of the game.

Since an n-person non-cooperative game may have more than one equilibrium point, and the payoffs to a certain player at two distinct equilibrium points are often different, no satisfactory simple notions of 'optimal strategy' and 'value' of the game are available. Nash's theorem guarantees the existence of an equilibrium point. However, proving the existence of an equilibrium point is far from being able to give a definition for a solution of an n-person non-cooperative game.

Intuitively, for the two players in the above example, the equilibrium situations

$$(X, Y) = ((1, 0), (1, 0)) \text{ and } ((0, 1), (0, 1))$$

are both more advantageous than the other equilibrium situation

$$(X, Y) = ((\tfrac{3}{5}, \tfrac{2}{5}), (\tfrac{2}{5}, \tfrac{3}{5})),$$

yet we cannot see how they could reach either of the former two equilibrium situations, since pre-play communication is explicitly forbidden.

Since this game is symmetric with respect to the two players, an ideal answer would be for both players to choose their strategy 1 simultaneously with probability $\tfrac{1}{2}$ and to choose their strategy 2 simultaneously with probability $\tfrac{1}{2}$. This would always lead to one of the two former equilibrium situations mentioned above, and the expected payoffs to both players would be $\tfrac{3}{2}$. Such a correlation of strategies, however, is beyond the range of the non-cooperative game. The expected payoffs $E_1 = E_2 = \tfrac{3}{2}$ are never possible in our non-cooperative game: the payoff point

$$(E_1, E_2) = (\tfrac{3}{2}, \tfrac{3}{2})$$

lies outside the shaded region V in Fig. 3.2. Acting in this way amounts practically to the players making a binding agreement before play that they both choose strategy 1 or both choose strategy 2, and to their redistribution of the joint payoffs after play — each gets one-half of the sum of their payoffs. All these actions are strictly prohibited in a non-cooperative game.

Example 3.4. The payoff matrices for the 'prisoners' dilemma' of Example 3.1 are

$$A = \begin{bmatrix} 8 & 0 \\ 10 & 2 \end{bmatrix}, \qquad B = \begin{bmatrix} 8 & 10 \\ 0 & 2 \end{bmatrix}. \tag{3.43}$$

We calculate Q, q, R, r by means of the formulae (3.20) and (3.25):

$$Q = 0, \qquad q = 2 > 0,$$
$$R = 0, \qquad r = 2 > 0.$$

It follows from (3.22) and (3.29) that equilibrium point (x, y) of the game satisfies the following relations:

$$x = 0, \qquad 0 \leqslant y \leqslant 1,$$
$$0 \leqslant x \leqslant 1, \qquad y = 0.$$

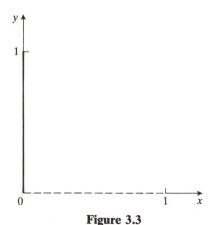

Figure 3.3

Hence the game has a unique equilibrium point, namely (see Fig. 3.3)

$$x = 0, \qquad y = 0,$$

or

$$(X, Y) = ((0, 1), (0, 1)).$$

Players 1 and 2 both choose strategy 2, i.e. both confess to be guilty, and both get the payoff 2, which amounts to a six-year sentence for both prisoners.

From the payoff matrices (3.43), it is obvious that this equilibrium situation is not the most advantageous situation to the players. If both players choose their first strategy, i.e. $x = 1$, $y = 1$, then both obtain the payoff 8, which means a one-year sentence for both prisoners. This is the most advantageous outcome for the two players. However, in the non-cooperative context it is hardly possible to arrive at this ideal situation in the absence of pre-play binding agreements.

It was mentioned in Example 3.1 that the outcomes of this problem can also be represented by the following payoff matrices:

$$A = \begin{bmatrix} -12 & -120 \\ -3 & -72 \end{bmatrix}, \qquad B = \begin{bmatrix} -12 & -3 \\ -120 & -72 \end{bmatrix}.$$

The calculation of the equilibrium points of this bimatrix game is left to the reader.

4

N-PERSON COOPERATIVE GAMES

4.1 Introduction

In an n-person non-cooperative game, any pre-play binding agreements regarding the correlation of strategies between two or more players, and any redistribution of payoffs among any group of players, are forbidden.

The n-person cooperative games to be discussed in this chapter are not restricted in these respects. Cooperation between players is permitted. Any group of players (called a coalition of players) can freely make pre-play binding agreements to correlate their strategies; the total payoff to a coalition of players may be divided in any way among these players. The latter condition is called the side payments condition. Thus we have the following elements to consider.

First, since certain players are to cooperate in some respects, a coalition will be formed by these players. This is an important matter in an n-person cooperative game. In a non-cooperative game, every player seeks to maximize his own individual payoff, and there is no problem of forming coalitions.

Second, after a coalition is formed by some players, this coalition as a whole will strive to gain as large a total payoff as possible. This payoff is a function of the coalition.

Third, the total payoff gained by each coalition will be divided among the members of the coalition. The share received by each player in a coalition will be represented by a number. This problem will be discussed in Section 4.3.

In an n-person cooperative game, the selection of strategies is no longer the main problem to be considered. The point to be emphasized is the formation of coalitions. We assume that, for an n-person non-cooperative game $[I, \{X_i\}, \{P_i\}]$, the set of players is $I = \{1, \ldots, n\}$. Let S be any subset of I, called a *coalition* of the game; and let $I \backslash S$ be the complementary set of S in I; this is another coalition of the game. Consider the zero-sum two-person matrix game which results when all players of S cooperate with each other on the one hand, and all players of $I \backslash S$ cooperate with each other on the other hand. The set of all joint combinations of pure strategies of the players in S is the set of pure strategies for the first player S of the zero-sum two-person matrix game. Similarly, all joint combinations of pure strategies of the players in $I \backslash S$ form the set of pure strategies for the second player $I \backslash S$ of the zero-sum two-person matrix game.

For every $S \subseteq I$, the above zero-sum two-person matrix game has a value. Denote this value by $v(S)$. We define further that

$$v(\varnothing) = 0.$$

$v(S)$ is called the *characteristic function* of the game. Characteristic function is an important element for the description of an n-person cooperative game.

Definition 4.1. Let $I = \{1, \ldots, n\}$. If $v(S)$ is a real-valued function defined on the set of all subsets (i.e. coalitions) S of I, satisfying

$$v(\varnothing) = 0, \tag{4.1}$$

$$v(I) \geqslant \sum_{i=1}^{n} v(\{i\}), \tag{4.2}$$

we say that $\Gamma \equiv [I, v]$ is an *n-person cooperative game*. $v(S)$ is called the characteristic function of the game Γ.

Throughout this chapter we assume that the amount received by each coalition S can be divided among its members in any way possible. This is the *side payments* condition already mentioned.

We give an example of a three-person cooperative game to explain how it can be expressed in the characteristic function form.

Example 4.1. Consider a three-person cooperative game in which each of players 1, 2 and 3 has two strategies A and B. The payoffs to players 1, 2, 3 are denoted by (a, b, c) in the following table:

	Players 1	2	3	Payoffs $(a,$	$b,$	$c)$
	A	A	A	$(1,$	$1,$	$0)$
	A	A	B	$(-3,$	$1,$	$2)$
	A	B	A	$(4,$	$-2,$	$2)$
Strategies	A	B	B	$(0,$	$1,$	$1)$
	B	A	A	$(1,$	$2,$	$-1)$
	B	A	B	$(2,$	$0,$	$-1)$
	B	B	A	$(3,$	$1,$	$-1)$
	B	B	B	$(2,$	$1,$	$-1)$

Let us compute the values of the characteristic function of this game. First, suppose players 1, 2 form a coalition $S = \{1, 2\}$. The payoff matrix of the zero-sum two-person matrix game between S and $I\backslash S = \{3\}$ is

		{3}	
		A	B
	AA	2	-2
$\{1, 2\}$	AB	2	1
	BA	3	2
	BB	4	3

This 4×2 matrix game has a saddle point, the value at the saddle point is 3. Hence

$$V(\{1, 2\}) = 3.$$

Next, let $S = \{3\}$. Then $I\backslash S = \{1, 2\}$. We have the following 2×4 matrix game:

		{1, 2}			
		AA	AB	BA	BB
$\{3\}$	A	0	2	-1	-1
	B	2	1	-1	-1

It is easily seen that

$$v(\{3\}) = -1.$$

Let $S = \{1, 3\}$. Then $I\backslash S = \{2\}$. We have the following 4×2 matrix game:

		{2}	
		A	B
	AA	1	6
$\{1, 3\}$	AB	-1	1
	BA	0	2
	BB	1	1

Obviously,

$$v(\{1, 3\}) = 1.$$

Let $S = \{2\}$. Then $I\backslash S = \{1, 3\}$. We have the following 2×4 matrix game:

		$\{1, 3\}$		
	AA	AB	BA	BB
A	1	1	2	0
B	-2	1	1	1

(with $\{2\}$ labelling the rows A, B.)

$S = \{2\}$ as the first player in this matrix game has the optimal mixed strategy $(\frac{3}{4}, \frac{1}{4})$. $I\backslash S = \{1, 3\}$ as the second player has the optimal mixed strategy $(\frac{1}{4}, 0, 0, \frac{3}{4})$. The value is $\frac{1}{4}$. Hence

$$v(\{2\}) = \tfrac{1}{4}.$$

Now let $S = \{2, 3\}$. Then $I\backslash S = \{1\}$. We have

	$\{1\}$	
	A	B
AA	1	1
AB	3	-1
BA	0	0
BB	2	0

(with $\{2, 3\}$ labelling the rows AA, AB, BA, BB.)

This matrix game has a saddle point at the upper right corner of the matrix. Hence

$$v(\{2, 3\}) = 1.$$

Finally, let $S = \{1\}$. Then $I\backslash S = \{2, 3\}$. We have

		$\{2, 3\}$		
	AA	AB	BA	BB
A	1	-3	4	0
B	1	2	3	2

(with $\{1\}$ labelling the rows A, B.)

Obviously,

$$v(\{1\}) = 1.$$

Thus the values of the characteristic function v are as follows:

$$v(\{1\}) = 1, \qquad v(\{2\}) = \tfrac{1}{4}, \qquad v(\{3\}) = -1,$$
$$v(\{1, 2\}) = 3, \quad v(\{1, 3\}) = 1, \quad v(\{2, 3\}) = 1,$$
$$v(\{1, 2, 3\}) = 4.$$

4.2   Properties of characteristic functions

Let $I = \{1, \ldots, n\}$. Consider the game $[I, \{X_i\}, \{P_i\}]$. If we denote by $v(S)$ the characteristic function of the cooperative game, then by definition,

$$v(S) = \max_{x \in X_S} \min_{y \in X_{I\backslash S}} \sum_{i \in S} E_i(x, y)$$

$$= \min_{y \in X_{I\backslash S}} \max_{x \in X_S} \sum_{i \in S} E_i(x, y), \qquad (4.3)$$

where X_S and $X_{I\backslash S}$ represent the sets of all correlated mixed strategies of members of coalitions S and $I\backslash S$, respectively. $E_i(x, y)$ is the expected payoff to player i for the pair of mixed strategies $x \in X_S$, $y \in X_{I\backslash S}$.

Theorem 4.1. *Let* $\Gamma \equiv [I, v]$ *be an n-person cooperative game. Then*

$$v(S \cup T) \geqslant v(S) + v(T) \qquad (4.4)$$

for all S, $T \subseteq I$, $S \cap T = \varnothing$.

Proof. The coalition S can obtain from its opponent $I\backslash S$ an amount $v(S)$ and no more. Similarly, the coalition T can obtain the amount $v(T)$ and no more. Hence the coalition $S \cup T$ can obtain the amount $v(S) + v(T)$ even if S and T fail to cooperate with each other. Since the maximum which the coalition $S \cup T$ can obtain under any condition is $v(S \cup T)$, this implies

$$v(S \cup T) \geqslant v(S) + v(T). \qquad \square$$

This simple proof is given in von Neumann and Morgensten (1953, p. 242). We agree with von Neumann that this conceptual proof is strictly equivalent to a formal mathematical proof. For a formal proof see Vorob'ev (1977, pp. 119–120).

From the above discussion we see that the characteristic function v of an *n*-person cooperative game $\Gamma \equiv [I, v]$ is a real-valued set function defined for all subsets of I and satisfying the conditions (4.1), (4.2) and (4.4).

Property (4.4) is called the *superadditivity* of the characteristic function v. If equality holds in (4.4), i.e. if for all S, $T \subseteq I$, $S \cap T = \varnothing$, we have

$$v(S \cup T) = v(S) + v(T), \qquad (4.5)$$

we say that v is *additive*.

An *n*-person cooperative game with additive characteristic function is called an *inessential* game. Inessential games are trivial and deserve none of our attention. This can be seen from the following theorem.

Theorem 4.2. *Let* $\Gamma \equiv [I, v]$ *be an n-person cooperative game. A necessary and sufficient condition for the characteristic function v to be additive is that*

$$v(I) = \sum_{i=1}^{n} v(\{i\}). \qquad (4.6)$$

Proof. The necessity of the condition is obvious. To prove the sufficiency of the condition, assume that (4.6) holds. Let S, $T \subseteq I$, $S \cap T = \emptyset$. Using superadditivity of v successively we have

$$v(I) = \sum_{i=1}^{n} v(\{i\})$$

$$= \sum_{i \in S} v(\{i\}) + \sum_{i \in T} v(\{i\}) + \sum_{i \in I \backslash S \cup T} v(\{i\})$$

$$\leqslant v(S) + v(T) + v(I \backslash S \cup T)$$

$$\leqslant v(S \cup T) + v(I \backslash S \cup T)$$

$$\leqslant v(I).$$

Hence

$$v(S) + v(T) = v(S \cup T). \qquad \square$$

We are interested chiefly in the n-person cooperative games for which the characteristic function v satisfies the condition

$$v(I) > \sum_{i=1}^{n} v(\{i\}). \qquad (4.7)$$

These games are called *essential* games.

4.3 Imputations

Every player of a cooperative game has a right to receive his share from the total payoff $v(I)$ available. A division of $v(I)$ among the players belonging to I can be represented by an n-dimensional vector

$$x = (x_1, \ldots, x_n), \qquad (4.8)$$

where x_i is the amount received by player i. This vector should satisfy the following two conditions:

$$x_i \geqslant v(\{i\}), \qquad i = 1, \ldots, n, \qquad (4.9)$$

$$\sum_{i=1}^{n} x_i = v(I). \qquad (4.10)$$

This vector x is called an *imputation*.

Relation (4.9) is the condition of *individual rationality*. If this condition is not satisfied, player i will certainly refuse to accept the distribution, since he can be sure of the amount $v(\{i\})$ even if he does not cooperate with any of the other players.

Relation (4.10) is the condition of *group rationality*. It is also called the condition of *Pareto optimality*. This condition is necessary. For, if

$$v(I) > \sum_{i=1}^{n} x_i,$$

then the players can form the grand coalition I to obtain the total payoff $v(I)$. Now that

$$v(I) - \sum_{i=1}^{n} x_i > 0,$$

each player i can receive an additional amount besides x_i; for example, every player of I can obtain an extra share of

$$\frac{1}{n}\left[v(I) - \sum_{i=1}^{n} x_i\right] > 0.$$

Therefore, such a distribution scheme will never be accepted by the players. On the other hand,

$$\sum_{i=1}^{n} x_i > v(I)$$

is impossible, since $v(I)$ is the greatest amount that the grand coalition I can obtain and the total distribution cannot exceed the total income. Thus the equality (4.10) should hold.

Any inessential game has one imputation only.

Theorem 4.3. *Any inessential cooperative game has one imputation*

$$x = (v(\{1\}), \ldots, v(\{n\})) \tag{4.11}$$

only.

Proof. Let $x = (x_1, \ldots, x_n)$ be an imputation of an inessential game $\Gamma \equiv [I, v]$. Assume that

$$x_i > v(\{i\})$$

for some i. Then by (4.9)

$$\sum_{i=1}^{n} x_i > \sum_{i=1}^{n} v(\{i\}).$$

Now the left-hand side of this inequality equals $v(I)$ by (4.10). Moreover, the right-hand side of this inequality also equals $v(I)$ by (4.6). Hence we have $v(I) > v(I)$, an absurdity. Therefore, for every i we have

$$x_i = v(\{i\}).$$

That is to say, the game has a unique imputation

$$x = (v(\{1\}), \ldots, v(\{n\})). \qquad \square$$

For an essential cooperative game $\Gamma \equiv [I, v]$, since

$$a = v(I) - \sum_{i=1}^{n} v(\{i\}) > 0,$$

there are an infinite number of ways to divide the quantity a into n non-negative real numbers $a_1, \ldots, a_n$ such that

$$\sum_{i=1}^{n} a_i = a.$$

It is easily verified that any vector of the form

$$x = (v(\{1\}) + a_1, \ldots, v(\{n\}) + a_n)$$

is an imputation of Γ. Therefore, Γ has infinitely many imputations.

4.4   Strategic equivalence and (0, 1) normalization

Cooperative games can be classified in such a way that any two games belonging to the same class have exactly the same strategic possibilities. We say that games of the same class are strategically equivalent; we also say that the corresponding characteristic functions are strategically equivalent.

After having classified all n-person cooperative games into strategically equivalent classes, it is desirable to pick from each class of characteristic functions in strategic equivalence a particularly simple representative. Thus the analysis of a class of games or characteristic functions reduces to the analysis of this simple representative.

Definition 4.2. Let $\Gamma \equiv [I, v]$ and $\Gamma' \equiv [I, v']$ be two n-person cooperative games defined on the same set of players $I = \{1, \ldots, n\}$. If there exist n numbers $a_1, \ldots, a_n$ and a positive constant c such that

$$v'(S) = cv(S) + \sum_{i \in S} a_i \qquad (4.12)$$

for all $S \subseteq I$, we say that Γ and Γ' are *strategically equivalent*. We also say that the characteristic functions v and v' are strategically equivalent.

We prove that the relation of strategic equivalence satisfies the three conditions of an equivalence relation, i.e. reflexivity, symmetry, and transitivity. We denote by the symbol '$v \sim v'$' the relation of strategic equivalence of v and v'.

(1) *Reflexivity*. Every characteristic function v is strategically equivalent to itself, i.e.

$$v \sim v.$$

To prove this property, we let $c = 1$, $a_i = 0$, $i = 1, \ldots, n$ in (4.12).

(2) *Symmetry*. If $v \sim v'$, then $v' \sim v$.
Let $v \sim v'$, i.e.

$$v'(S) = cv(S) + \sum_{i \in S} a_i.$$

Solving for $v(S)$, we have

$$v(S) = \frac{1}{c} v'(S) + \sum_{i \in S} \left(-\frac{a_i}{c} \right).$$

Since $1/c > 0$, and $-(a_i/c)$ $(i = 1, \ldots, n)$ are all constants, the above equation is the definition of $v' \sim v$.

(3) *Transitivity*. If $v \sim v'$, $v' \sim v''$, then $v \sim v''$. Since $v \sim v'$, we have

$$v'(S) = cv(S) + \sum_{i \in S} a_i \quad (c > 0). \tag{4.12}$$

Since $v' \sim v''$, we have

$$v''(S) = c'v'(S) + \sum_{i \in S} a_i' \quad (c' > 0). \tag{4.13}$$

Substituting (4.12) into (4.13) we obtain

$$v''(S) = c'cv(S) + \sum_{i \in S} (a_i' + c'a_i).$$

$c'c$ is a positive constant. $a_i' + c'a_i$, $i = 1, \ldots, n$, are n real numbers. Therefore, $v \sim v''$.

The concept of strategic equivalence enables us to study only a single game from each class. We wish to choose from each class the simplest game as the representative. For this purpose, let us first introduce the following definition.

Definition 4.3. If the characteristic function v of an n-person cooperative game $\Gamma \equiv [I, v]$ satisfies the conditions

$$v(\{i\}) = 0, \qquad i = 1, \ldots, n, \tag{4.14}$$

$$v(I) = 1, \tag{4.15}$$

we say that Γ or v is in $(0, 1)$ *normalization*.

Theorem 4.4. *Every essential n-person cooperative game $\Gamma \equiv [I, v]$ is strategically equivalent to a unique game in $(0, 1)$ normalization.*

Proof. Let v' be a characteristic function in $(0, 1)$ normalization. In order to show that v is strategically equivalent to v', we prove that there exist n constants a_i, $i = 1, \ldots, n$, and a positive constant c satisfying

$$v'(\{i\}) = cv(\{i\}) + a_i = 0, \qquad i = 1, \ldots, n, \tag{4.16}$$

$$v'(I) = cv(I) + \sum_{i=1}^{n} a_i = 1. \tag{4.17}$$

Summing the n equations in (4.16) we obtain

$$\sum_{i=1}^{n} a_i = -c \sum_{i=1}^{n} v(\{i\}).$$

Substituting into (4.17) gives

$$cv(I) - c \sum_{i=1}^{n} v(\{i\}) = 1.$$

Since by hypothesis the game is essential, we have

$$v(I) - \sum_{i=1}^{n} v(\{i\}) > 0.$$

Hence

$$c = 1 / \left[v(I) - \sum_{i=1}^{n} v(\{i\}) \right] > 0. \tag{4.18}$$

Substituting (4.18) for the c in (4.16), we get

$$a_i = -cv(\{i\}) = -v(\{i\}) / \left[v(I) - \sum_{i=1}^{n} v(\{i\}) \right]. \tag{4.19}$$

Equations (4.18) and (4.19) constitute the unique solution for the system (4.16), (4.17). Therefore, the $(0, 1)$ normalized characteristic function v' strategically equivalent to v is uniquely determined. □

For an n-person cooperative game in $(0, 1)$ normalization, the defining conditions (4.9), (4.10) of an imputation

$$x = (x_1, \ldots, x_n) \tag{4.8}$$

become

$$x_i \geqslant 0, \qquad i = 1, \ldots, n, \tag{4.20}$$

$$\sum_{i=1}^{n} x_i = 1. \tag{4.21}$$

The set of all imputations is an $(n-1)$-dimensional simplex in the n-dimensional space.

It follows from Theorem 4.4 that, for a fixed I, every class of games in strategic equivalence can be represented by its $(0, 1)$ normalized game which possesses all the common properties of all members of the class.

Example 4.2. Suppose the values of the characteristic function v of a six-person cooperative game are

$$
\begin{aligned}
v(S) &= 1, &&\text{if } |S| = 1, \\
v(S) &= 3, &&\text{if } |S| = 2, \\
v(S) &= 5, &&\text{if } |S| = 3, \\
v(S) &= 7, &&\text{if } |S| = 4, \\
v(S) &= 10, &&\text{if } |S| = 5, \\
v(I) &= 12,
\end{aligned}
$$

where $|S|$ is the number of players in the coalition S. Let us write the characteristic function in $(0, 1)$ normalization form.

Calculating c and a_i according to the formulae (4.18) and (4.19), we have

$$c = 1 \Big/ \Big[v(I) - \sum_{i=1}^{6} v(\{i\}) \Big] = 1/(12 - 6) = \tfrac{1}{6};$$

$$a_i = -cv(\{i\}) = -\tfrac{1}{6} \cdot 1 = -\tfrac{1}{6}, \qquad i = 1, \ldots, 6.$$

The definition (4.12) gives

$$
\begin{aligned}
v'(\{i\}) &= 0, &&i = 1, \ldots, 6, \\
v'(S) &= \tfrac{3}{6} - \tfrac{2}{6} = \tfrac{1}{6}, &&\text{if } |S| = 2, \\
v'(S) &= \tfrac{5}{6} - \tfrac{3}{6} = \tfrac{2}{6}, &&\text{if } |S| = 3, \\
v'(S) &= \tfrac{7}{6} - \tfrac{4}{6} = \tfrac{3}{6}, &&\text{if } |S| = 4, \\
v'(S) &= \tfrac{10}{6} - \tfrac{5}{6} = \tfrac{5}{6}, &&\text{if } |S| = 5, \\
v'(I) &= 1.
\end{aligned}
$$

In the original game, any imputation

$$x = (x_1, \ldots, x_6)$$

satisfies the conditions

$$x_i \geqslant 1, \qquad i = 1, \ldots, 6, \qquad x_1 + \cdots + x_6 = 12.$$

For the strategically equivalent $(0, 1)$ normalized game, the imputations are the set of all vectors

$$x' = (x'_1, \ldots, x'_6)$$

satisfying

$$x'_i \geqslant 0, \qquad i = 1, \ldots, 6, \qquad x'_1 + \cdots + x'_6 = 1.$$

For the n-person cooperative games, as well as the $(0, 1)$ normalization, sometimes the $(-1, 0)$ normalization is used. This is the case in which the characteristic function v' satisfies the conditions

$$v'(\{i\}) = -1, \qquad i = 1, \ldots, n, \tag{4.22}$$

$$v'(I) = 0. \tag{4.23}$$

It is easy to see that the transformation formula from a $(0, 1)$ normalized characteristic function v to the corresponding $(-1, 0)$ normalized characteristic function v' is

$$v'(S) = nv(S) - |S| \tag{4.24}$$

for every $S \subseteq I$.

For the game in Example 4.2 above, if we transform the characteristic function v' in $(0, 1)$ normalization into a characteristic function v'' in $(-1, 0)$ normalization, the result is

$$\begin{aligned}
v''(\{i\}) &= -1, & i &= 1, \ldots, 6, \\
v''(S) &= -1, & \text{if } |S| &= 2, 3, 4, \\
v''(S) &= 0, & \text{if } |S| &= 5, \\
v''(I) &= 0.
\end{aligned}$$

Theorem 4.4 provides us with a method to reduce any essential cooperative game to its $(0, 1)$ normalized form. Now let us consider inessential cooperative games.

Example 4.3. Let $\Gamma \equiv [I, v]$ be an inessential n-person cooperative game. Since the characteristic function v of an inessential game is additive, we have

$$v(S) = \sum_{i \in S} v(\{i\}) \tag{4.25}$$

for all $s \subseteq I$. For every $S \subseteq I$, let

$$v'(S) = v(S) + \sum_{i \in S} [-v(\{i\})].$$

Comparing the above equation with the definition (4.12) of strategic equivalence, we have

$$c = 1 > 0, \qquad a_i = -v(\{i\}), \qquad i = 1, \ldots, n.$$

Hence, $v \sim v'$. However, $v'(S) \equiv 0$. Therefore, all inessential games are strategically equivalent to the game whose characteristic function is identically zero.

4.5 Two-person cooperative games

We discuss in this section properties of the characteristic functions of two-person cooperative games. Two-person cooperative games can be divided into two classes: constant-sum and non-constant-sum games.

A constant-sum n-person game is a game for which there exists a constant k such that

$$\sum_{i=1}^{n} P_i(s) = k$$

for every situation s of the game. If we choose n constants k_i, $i = 1, \ldots, n$, satisfying

$$\sum_{i=1}^{n} k_i = k$$

and let

$$P_i'(s) = P_i(s) - k_i$$

for every situation s, then

$$\sum_{i=1}^{n} P_i'(s) = 0$$

for every s. Thus a constant-sum game is equivalent to a zero-sum game strategically. The two games differ in that under the same strategy situation there is a constant difference k_i between the payoffs in the two games for the player i, $i = 1, \ldots, n$.

For n-person cooperative games in characteristic function form, constant-sum and zero-sum are also equivalent notions. In the first place, we have the following property for the characteristic functions of constant-sum games.

Theorem 4.5. *Let* $\Gamma \equiv [I, v]$ *be a constant-sum n-person cooperative game. Then*

$$v(S) + v(I \backslash S) = v(I) \qquad (4.26)$$

for every $S \subseteq I$.

Proof. We have

$$v(I) = \sum_{i=1}^{n} P_i(s) = k = \sum_{i=1}^{n} E_i(x),$$

where k is a constant, P_i is the payoff function of player i, E_i is his expected payoff. Hence, for every coalition $S \subseteq I$, by the definition (4.3) of a characteristic function,

$$\begin{aligned}
v(S) &= \max_{x \in X_S} \min_{y \in X_{I \backslash S}} \sum_{i \in S} E_i(x, y) \\
&= \max_{x \in X_S} \min_{y \in X_{I \backslash S}} \left[k - \sum_{i \in I \backslash S} E_i(x, y) \right] \\
&= \max_{x \in X_S} \left(k + \min_{y \in X_{I \backslash S}} \left[-\sum_{i \in I \backslash S} E_i(x, y) \right] \right) \\
&= \max_{x \in X_S} \left[k - \max_{y \in X_{I \backslash S}} \sum_{i \in I \backslash S} E_i(x, y) \right] \\
&= k - \min_{x \in X_S} \max_{y \in X_{I \backslash S}} \sum_{i \in I \backslash S} E_i(x, y) \\
&= k - \max_{y \in X_{I \backslash S}} \min_{x \in X_S} \sum_{i \in I \backslash S} E_i(x, y) \\
&= v(I) - v(I \backslash S).
\end{aligned}$$

From this (4.26) follows.   □

This property of constant-sum games is called the *complementarity* of characteristic functions.

We now return to the discussion of two-person cooperative games. For the constant-sum two-person cooperative game, if the game is essential, then by Theorem 4.4 it is strategically equivalent to a unique $(0, 1)$ normalized game v. The values of v are

$$v(\{1\}) = v(\{2\}) = 0, \qquad v(\{1, 2\}) = 1.$$

However, it follows from Theorem 4.5 that

$$v(\{1\}) = v(\{1, 2\}) - v(\{2\}) = 1 - 0 = 1,$$

which contradicts $v(\{1\}) = 0$. Thus we see that all constant-sum two-person cooperative games are inessential.

Now let us consider non-constant-sum two-person cooperative games.

Example 4.4. We reconsider the 2×2 bimatrix game of Example 3.3 ('the battle of the sexes'):

$$A = \begin{bmatrix} 2 & -1 \\ -1 & 1 \end{bmatrix}, \qquad B = \begin{bmatrix} 1 & -1 \\ -1 & 2 \end{bmatrix}.$$

We have seen that, as a non-cooperative game, since players 1 and 2 can only choose their mixed strategies independently, the payoff set of this game is the region V in Fig. 3.2 (p. 87).

Now, if we consider the game represented by the payoff matrices A and B as a two-person cooperative game, then the mixed strategies of the two players can be combined in any way possible so as to form the joint mixed strategies. For example, the players can make a pre-play binding agreement that they choose strategy 1 simultaneously, or they choose strategy 2 simultaneously. To do this, they simply toss a coin. If the coin shows heads up, they both choose strategy 1; if it shows tails up, they both choose strategy 2.

If the players are permitted to use all possible correlated mixed strategies, then the set of all payoff pairs (E_1, E_2) is the region V' in Fig. 4.1, which is the convex hull of the three points $(-1, -1)$, $(1, 2)$, $(2, 1)$.

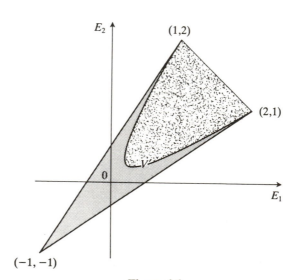

Figure 4.1

It is not difficult to calculate the values of the characteristic function v of this two-person cooperative game. They are

$$v(\{1\}) = v(\{2\}) = \tfrac{1}{5}, \qquad v(\{1, 2\}) = 3.$$

Example 4.5. The payoff matrices of the 'prisoners' dilemma' of Example 3.4 are

$$A = \begin{bmatrix} 8 & 0 \\ 10 & 2 \end{bmatrix}, \qquad B = \begin{bmatrix} 8 & 10 \\ 0 & 2 \end{bmatrix}.$$

As a two-person non-cooperative game, there is only one equilibrium point $x = 0$, $y = 0$, i.e. both players choose their second pure strategies. The situation will be quite different if we regard A, B as the payoff matrices of a two-person cooperative game. The two players can then make pre-play binding agreement regarding the choices of certain pure or mixed strategies. They can also share the total payoff they receive. We now calculate the characteristic function of this two-person cooperative game according to the definition (4.3).

Using the method of Example 4.1, we denote by I, II the first and second pure strategies of the two players and write down the following table:

		{2}	
		I	II
{1}	I	8	0
	II	10	2

Obviously, $v(\{1\}) = 2$.

Next, we have

		{1}	
		I	II
{2}	I	8	0
	II	10	2

It should be noted that we are now regarding $\{2\}$ as the first player S and $\{1\}$ as the second player $I\backslash S$; hence the payoff matrix here is the transpose of the original matrix B. Therefore, $v(\{2\}) = 2$. We obtain the values of the characteristic function v:

$$v(\{1\}) = v(\{2\}) = 2, \qquad v(\{1, 2\}) = 16.$$

Finally, we point out that any essential (and hence non-constant-sum) two-person cooperative game is strategically equivalent to a unique $(0, 1)$ normalized game whose characteristic function v has the values

$$v(\{1\}) = v(\{2\}) = 0, \qquad v(\{1, 2\}) = 1.$$

The two-person cooperative games in the above two examples can both be reduced to this strategically equivalent game.

4.6 Domination of imputations. Three-person cooperative games

Definition 4.4. Let $x = (x_1, \ldots, x_n)$ and $y = (y_1, \ldots, y_n)$ be two imputations of an n-person cooperative game $\Gamma \equiv [I, v]$, and let S be a non-empty subset of I. If

$$v(S) \geqslant \sum_{i \in S} y_i, \tag{4.27}$$

and

$$y_i > x_i, \qquad i \in S, \tag{4.28}$$

we say that the imputation y *dominates* the imputation x with respect to the coalition S, or x is dominated by y with respect to S. In symbolic form:

$$y \succ_S x. \tag{4.29}$$

Relation (4.27) is called the *effectiveness* or feasibility condition. We also say that the set S is *effective* for the imputation y. If this condition is not satisfied, i.e. if

$$v(S) < \sum_{i \in S} y_i,$$

then the members of S may not be able to obtain what they are offered by the imputation y, and so it makes no sense for the coalition S to consider any domination involving y.

The condition (4.28) states that the members of S all prefer imputation y to imputation x, since the former offers each member i of S more than the latter does.

The relation of domination of imputations with respect to a coalition S is transitive. That is to say, if $z \succ_S y$, $y \succ_S x$, then $z \succ_S x$.

Proof. Since $z \succ_S y$, we have

$$v(S) \geqslant \sum_{i \in S} z_i, \tag{4.30}$$

$$z_i > y_i, \qquad i \in S. \tag{4.31}$$

Since $y >_S x$, we have

$$v(S) \geq \sum_{i \in S} y_i, \tag{4.32}$$

$$y_i > x_i, \qquad i \in S. \tag{4.33}$$

Combining (4.30), (4.31), (4.33) we obtain

$$v(S) \geq \sum_{i \in S} z_i,$$

$$z_i > x_i, \qquad i \in S.$$

Hence $z >_S x$.

   Domination with respect to a one-person coalition cannot occur. For, suppose $y >_{\{i\}} x$, then by definition

$$v(\{i\}) \geq y_i, \qquad y_i > x_i.$$

Thus

$$v(\{i\}) \geq y_i > x_i,$$

which violates condition (4.9) for the definition of an imputation.

   Domination with respect to the grand coalition I is also not possible. For, suppose $y >_I x$, then

$$v(I) \geq \sum_{i \in I} y_i,$$

$$y_i > x_i, \qquad i \in I.$$

Thus

$$\sum_{i=1}^{n} y_i > \sum_{i=1}^{n} x_i = v(I),$$

which violates condition (4.10).

Definition 4.5. If there exists a non-empty coalition $S \subset I$ such that

$$y >_S x, \tag{4.34}$$

we say that the imputation y dominates the imputation x. In symbols,

$$y > x. \tag{4.35}$$

   The relation of general domination is not necessarily transitive. For example, let the values of the characteristic function v of a three-person cooperative game be

$$v(\{i\}) = 0, \qquad i = 1, 2, 3,$$

$$v(\{1, 2\}) = v(\{1, 3\}) = v(\{2, 3\}) = v(I) = 10.$$

Consider the following three imputations:
$$z = (0, 5, 5), \qquad y = (6, 4, 0), \qquad x = (4, 0, 6).$$
We have
$$z > y, \qquad y > x,$$
but x is not dominated by z.

Definition 4.6. Suppose $S \subset I$, y is an imputation. Define
$$\text{Dom}_S\, y = \{x : y >_S x\}. \tag{4.36}$$
It is called the *dominion* of the imputation y with respect to the coalition S. Define
$$\text{Dom}\, y = \bigcup_{\substack{S \subset I \\ S \neq \varnothing}} \text{Dom}_S\, y. \tag{4.37}$$
It is called the dominion of y.

Let X be the set of all imputations. For any set $A \subset X$, define
$$\text{Dom}_S\, A = \bigcup_{y \in A} \text{Dom}_S\, y. \tag{4.38}$$
It is called the dominion of the imputation set A with respect to the coalition S. Define
$$\text{Dom}\, A = \bigcup_{\substack{S \subset I \\ S \neq \varnothing}} \text{Dom}_S\, A. \tag{4.39}$$
It is called the dominion of A.

We shall consider the domination of imputations in a three-person cooperative game. Since an inessential cooperative game has only one imputation, and all inessential games are strategically equivalent to the game whose characteristic function is identically zero (see Example 4.3), no domination of imputations need be considered. We assume that the three-person cooperative game to be considered is essential. Moreover, suppose that the characteristic function v is in $(0, 1)$ normalization.

For a three-person cooperative game, any point x of the imputation set X can be represented by a point with barycentric coordinates (x_1, x_2, x_3) in an equilateral triangle with unit perpendicular bisectors of sides, where x_1, x_2, x_3 satisfy
$$x_i \geqslant 0, \qquad i = 1, 2, 3, \qquad x_1 + x_2 + x_3 = 1.$$
This set is the S_3 of Section 1.12.

First, for an essential constant-sum three-person cooperative game

$\Gamma \equiv [I, v]$, we have

$$v(\{i\}) = 0, \qquad i = 1, 2, 3,$$

$$v(\{1, 2\}) = v(\{1, 3\}) = v(\{2, 3\} = v(I) = 1.$$

These values are obtained directly from the complementarity relation (4.26) of Theorem 4.5.

Let $y = (y_1, y_2, y_3)$ be a point of X. Consider first the coalition $\{1, 2\}$. We know that the definition of

$$y \succ_{\{1,2\}} x$$

is

$$v(\{1, 2\}) = 1 \geqslant y_1 + y_2, \qquad y_1 > x_1, \qquad y_2 > x_2.$$

Hence, in Fig. 4.2, the dominion of y with respect to the coalition $\{1, 2\}$ is

$$\text{Dom}_{\{1,2\}} \, y = yb3c, \quad \text{excluding } yb \text{ and } cy.$$

Simularly, we have

$$\text{Dom}_{\{1,3\}} \, y = yf2a, \quad \text{excluding } yf \text{ and } ay;$$

$$\text{Dom}_{\{2,3\}} \, y = yd1e, \quad \text{excluding } yd \text{ and } ey.$$

By (4.37), the dominion of y is

$$\text{Dom} \, y = (\text{Dom}_{\{1,2\}} \, y) \cup (\text{Dom}_{\{1,3\}} \, y) \cup (\text{Dom}_{\{2,3\}} \, y),$$

which is shown in Fig. 4.2 as the three shaded regions. The imputations

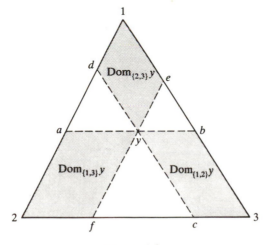

Figure 4.2

in the three triangles including the sides are not dominated by the imputation y. These imputations form the area which is equal to

$$X \backslash \text{Dom } y.$$

Now let us consider the domination of imputations for a non-constant-sum three-person cooperative game. The values of the characteristic function v of a general (non-constant-sum) three-person cooperative game $\Gamma \equiv [I, v]$ in $(0, 1)$ normalization are

$$v(\{i\}) = 0, \qquad i = 1, 2, 3,$$

$$v(I) = 1,$$

$$v(\{1, 2\}) = c_3, \qquad v(\{1, 3\}) = c_2, \qquad v(\{2, 3\}) = c_1,$$

where c_1, c_2, c_3 are constants in the closed interval $[0, 1]$.

Since no domination of imputations is possible with respect to either a one-person coalition or the grand coalition I, it is sufficient to consider the three two-person coalitions $\{1, 2\}$, $\{1, 3\}$, $\{2, 3\}$. Consider first the domination with respect to the coalition $\{1, 2\}$

$$y \succ_{\{1, 2\}} x, \tag{4.40}$$

where $x = (x_1, \ldots, x_n)$, $y = (y_1, \ldots, y_n)$ are imputations in X. By definition,

$$v(\{1, 2\}) = c_3 \geqslant y_1 + y_2, \tag{4.41}$$

$$y_1 > x_1, \qquad y_2 > x_2. \tag{4.42}$$

The effectiveness condition can be written in the form

$$y_3 \geqslant 1 - c_3. \tag{4.43}$$

This inequality means that the point y should lie on or to the right of the line $y_3 = 1 - c_3$. See Fig. 4.3.

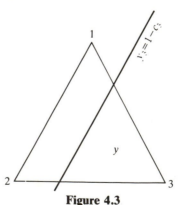

Figure 4.3

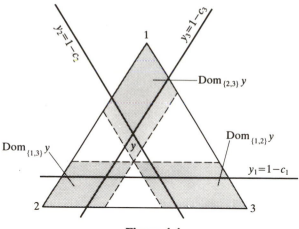

Figure 4.4

Similarly, from the dominations with respect to the coalitions $\{1,3\}$ and $\{2,3\}$, i.e. from

$$y \succ_{\{1,3\}} x, \qquad y \succ_{\{2,3\}} x,$$

we have

$$y_2 \geqslant 1 - c_2, \tag{4.44}$$

$$y_1 \geqslant 1 - c_1, \tag{4.45}$$

respectively. The inequality (4.44) means that the point y should lie on or to the left of the line $y_2 = 1 - c_2$; (4.45) means that y should lie on or above the line $y_1 = 1 - c_1$. See Fig. 4.4.

If an imputation y satisfies the conditions (4.43)–(4.45), then its dominion is the shaded region shown in Fig. 4.4. It consists of three parallelograms. This situation is just the same as in Fig. 4.2. If y satisfies (4.45) but does not satisfy (4.43) and (4.44), then Dom y is as shown in Fig. 4.5.

Again, if the values of c_1, c_2, c_3 are such that the three lines

$$y_1 = 1 - c_2, \qquad y_2 = 1 - c_2, \qquad y_3 = 1 - c_3$$

and the point y are situated as shown in Fig. 4.6, then the dominion of y consists of the two shaded parallelograms in the figure. The figures for the other cases can be obtained in a similar way. It is really unnecessary to enumerate all the possibilities.

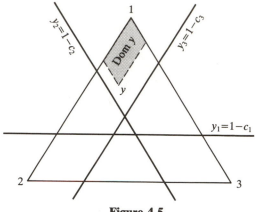

Figure 4.5

4.7 The core of a cooperative game

Beginning with this section, we shall deal with some of the solution concepts for n-person cooperative games. Let $\Gamma \equiv [I, v]$ be an n-person cooperative game, where $I = \{1, \ldots, n\}$ and the characteristic function v is a real-valued function defined on the class of all subsets of I. Intuitively speaking, we wish to single out from the set X of all imputations some imputations which the players of the various coalitions are willing to

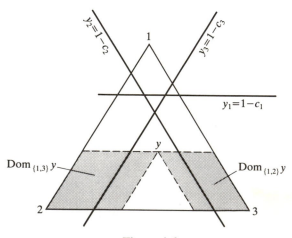

Figure 4.6

accept. If there exists a unique imputation $x \in X$ which will satisfy every player i, this imputation can serve as a solution for the game under consideration. For example, in the essential three-person cooperative game discussed in the preceding section, if there is an imputation x which dominates any other imputation and which is not dominated by any other imputation, that is to say, if no coalition can recommend a more advantageous (feasible) imputation, then this imputation x is an ideal distribution scheme and should be welcome to all players of the game. This imputation can be regarded as a solution of the game. Unfortunately, such an imputation does not exist in general.

Thus, the first thing for us to do is to seek for a set of imputations satisfying certain conditions, rather than a single imputation. We shall consider first those imputations which are not dominated by any other imputation.

The following notation will be used. Let $S \neq \varnothing$ be a coalition of players in I, and let $x = (x_1, \ldots, x_n)$ be an imputation. We denote

$$x(S) = \sum_{i \in S} x_i, \qquad \text{for} \quad S \neq \varnothing. \tag{4.46}$$

For the empty set $\varnothing$, we define

$$x(\varnothing) = 0. \tag{4.47}$$

Definition 4.7. Let $\Gamma \equiv [I, v]$ be an n-person cooperative game. The set $C(\Gamma)$ of imputations x satisfying

$$v(S) \leqslant x(S) \tag{4.48}$$

for all $S \subseteq I$ is called the *core* of Γ. That is

$$C(\Gamma) = \{x \colon x \in X; v(S) - x(S) \leqslant 0, S \subseteq I\}. \tag{4.49}$$

Relation (4.48) means that, for every coalition $S \subseteq I$, an imputation x of $C(\Gamma)$ offers to S an amount which is at least the amount it can realize by its own right. Hence x is an imputation acceptable to all S.

The following theorem shows that the imputations of the core are the imputations undominated by any other imputations.

Theorem 4.6. *Let* $\Gamma \equiv [I, v]$ *be an n-person cooperative game, and let* $C(\Gamma)$ *be the core of* Γ. *If* $x = (x_1, \ldots, x_n)$ *is an imputation, then a necessary and sufficient condition for* $x \in C(\Gamma)$ *is that x be undominated.*

Proof. *Necessity.* We want to prove that if $x \in C(\Gamma)$, then x is undominated. Suppose that x is dominated by some imputation $y = (y_1, \ldots, y_n)$, i.e. $y > x$. Then, by the definition of domination of imputations, there exists some $S \subset I$ such that

$$v(S) \geqslant y(S) > x(S).$$

Hence $x \notin C(\Gamma)$.

Sufficiency. We now prove that if x is undominated, then $x \in C(\Gamma)$. We may assume that the characteristic function v is in $(0, 1)$ normalization. Suppose that $x \notin C(\Gamma)$. Then there exists a coalition S such that

$$v(S) > x(S).$$

S cannot be a one-person coalition, nor can it be the grand coalition I. Hence we have

$$
\begin{aligned}
x(I \backslash S) &= x(I) - x(S) \\
&= v(I) - x(S) \\
&\geqslant v(S) - x(S) > 0.
\end{aligned}
$$

We construct an imputation $y = (y_1, \ldots, y_n)$ such that

$$y >_S x.$$

For this purpose, let ε be a number satisfying

$$0 < \varepsilon < \frac{1}{|S|} [v(S) - x(S)], \tag{4.50}$$

and let

$$
y_i =
\begin{cases}
x_i + \varepsilon, & i \in S, \\
\dfrac{1}{n - |S|} [x(I \backslash S) - |S| \, \varepsilon], & i \notin S.
\end{cases}
$$

It is easily verified that

$$y_i > 0, \qquad i = 1, \ldots, n$$

[by (4.50), $v(S) - x(S) > |S| \, \varepsilon$. Hence $i \notin S$ implies

$$
\begin{aligned}
x(I \backslash S) - |S| \, \varepsilon &= x(I) - x(S) - |S| \, \varepsilon \\
&= v(I) - x(S) - |S| \, \varepsilon \\
&\geqslant v(S) - x(S) - |S| \, \varepsilon > 0],
\end{aligned}
$$

and

$$y(I) = v(I).$$

Thus y is an imputation. Moreover,

$$
\begin{aligned}
v(S) &> x(S) + |S| \, \varepsilon = y(S), \\
y_i &> x_i, \qquad i \in S.
\end{aligned}
$$

This proves $y >_S x$ and the sufficiency of the condition is proved.   □

It is known that no domination of imputations is possible with respect to either a one-person coalition or the grand coalition I. Hence for an n-person cooperative game to have a non-empty core it is necessary that $n \geqslant 3$.

When $n \geqslant 3$, the characteristic function v of any inessential game $\Gamma \equiv [I, v]$ is additive and has only one imputation, namely

$$x = (v(\{1\}), \ldots, v(\{n\})).$$

Thus this imputation may also be considered as the unique element in the core of Γ.

For $n \geqslant 3$, the essential n-person cooperative games can be divided into two classes: constant-sum games and non-constant-sum games. Regarding the former, we have the following theorem.

Theorem 4.7. *Let* $\Gamma \equiv [I, v]$ *be an essential constant-sum n-person cooperative game. Then the core of the game is* $C(\Gamma) = \varnothing$.

Proof. Assume that the core is not empty, and let $x \in C(\Gamma)$. Then by definition (4.49), for every i we have

$$v(\{i\}) \leqslant x_i, \tag{4.51}$$

$$v(I \backslash i) \leqslant x(I \backslash i). \tag{4.52}$$

Now sum the two inequalities. Since the game is constant-sum, the sum of the two left-hand terms is $v(I)$. On the other hand, by the definition of an imputation (4.10), the two right-hand terms also add up to $v(I)$. Hence equality holds in (4.51) for all i, i.e.

$$v(\{i\}) = x_i, \qquad i = 1, \ldots, n. \tag{4.53}$$

It follows from (4.53) that the game is inessential. Therefore, $C(\Gamma) = \varnothing$.

$\square$

This theorem states that if an essential n-person cooperative game is constant-sum (for $n \geqslant 3$), then its core is an empty set.

In the remaining part of this section we give a few examples to illustrate the cores of the essential non-constant-sum three-person cooperative games.

Example 4.6. Suppose the values of the characteristic function v of a three-person cooperative game Γ are

$$v(\{i\}) = 0, \qquad i = 1, 2, 3,$$

$$v(\{1, 2\}) = \tfrac{2}{3}, \qquad v(\{1, 3\}) = \tfrac{7}{12}, \qquad v(\{2, 3\}) = \tfrac{1}{2},$$

$$v(\{1, 2, 3\}) = 1.$$

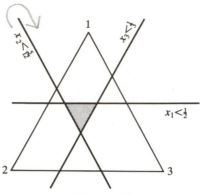

Figure 4.7

By the definition of the core, the imputation $x = (x_1, x_2, x_3) \in C(\Gamma)$ if and only if

$$v(\{1\}) = 0 \leq x_1,$$
$$v(\{2\}) = 0 \leq x_2,$$
$$v(\{3\}) = 0 \leq x_3,$$
$$v(\{1, 2\}) = \tfrac{2}{3} \leq x_1 + x_2,$$
$$v(\{1, 3\}) = \tfrac{7}{12} \leq x_1 + x_3,$$
$$v(\{2, 3\}) = \tfrac{1}{2} \leq x_2 + x_3.$$

By means of the condition of group rationality (4.10), the last three of the above inequalities reduce to

$$x_3 \leq \tfrac{1}{3}, \qquad x_2 \leq \tfrac{5}{12}, \qquad x_1 \leq \tfrac{1}{2}.$$

Therefore, the core of Γ is the shaded area in Fig. 4.7. It is a triangle including its sides.

Example 4.7. Suppose the values of the characteristic function v of a three-person cooperative game Γ are

$$v(\{i\}) = 0, \qquad i = 1, 2, 3,$$
$$v(\{1, 2\}) = \tfrac{1}{3}, \qquad v(\{1, 3\}) = \tfrac{1}{6}, \qquad v(\{2, 3\}) = \tfrac{5}{6},$$
$$v(\{1, 2, 3\}) = 1.$$

$x \in C(\Gamma)$ if and only if

$$x_i \geq 0, \qquad i = 1, 2, 3,$$
$$v(\{1, 2\}) = \tfrac{1}{3} \leq x_1 + x_2,$$
$$v(\{1, 3\}) = \tfrac{1}{6} \leq x_1 + x_3,$$
$$v(\{2, 3\}) = \tfrac{5}{6} \leq x_2 + x_3.$$

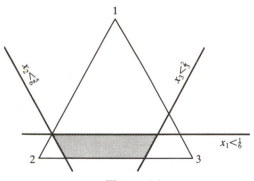

Figure 4.8

The last three inequalities reduce to

$$x_3 \leqslant \tfrac{2}{3}, \qquad x_2 \leqslant \tfrac{5}{6}, \qquad x_1 \leqslant \tfrac{1}{6}.$$

Therefore, the core of Γ is the shaded region in Fig. 4.8. It is a quadrangle.

From these two examples it should be clear that, if an essential non-constant-sum three-person cooperative game has a non-empty core, then it can be a point, a line segment, a triangle, a quadrangle, a pentagon, or a hexagon.

On the other hand, an essential non-constant-sum three-person cooperative game may not have a non-empty core. The core can be an empty set. In fact, for the non-constant-sum three-person cooperative games in $(0, 1)$ normalization, the core is empty if

$$v(\{1, 2\}) + v(\{1, 3\}) + v(\{2, 3\}) > 2. \tag{4.54}$$

Example 4.8. Suppose the values of the characteristic function v of a three-person cooperative game are

$$v(\{i\}) = 0, \qquad i = 1, 2, 3,$$
$$v(\{1, 2\}) = \tfrac{2}{3}, \qquad v(\{1, 3\}) = \tfrac{2}{3}, \qquad v(\{2, 3\}) = \tfrac{3}{4},$$
$$v(\{1, 2, 3\}) = 1.$$

This game has an empty core.

4.8   Stable sets of cooperative games

The concept of the core has been discussed in the preceding section. If the core is non-empty for an n-person cooperative game, then the imputations of the core are the imputations undominated by any other imputations. Without doubt the core is an important concept of the cooperative games. However, if we endeavour to treat the core as a solution concept of cooperative games, we encounter an insurmountable difficulty: many games have empty cores.

von Neumann and Morgenstern (1953) proposed a concept of solution for a cooperative game. A solution is defined as a set V of imputations such that no imputation of V is dominated by any other imputation of V, and any imputation not in V is dominated by some imputation of V.

Definition 4.8. Let X be the set of all imputations of an n-person cooperative game $\Gamma \equiv [I, v]$. If $V \subseteq X$ is a set of imputations satisfying the conditions:

(1)   for any $x \in V$ and $y \in V$, $x \not\succ y$,   (4.55)

(2)   if $w \notin V$, then there exists $z \in V$

         such that $z \succ w$,   (4.56)

then the set V is called a *stable set* of the game Γ, or a vN–M *solution* (von Neumann–Morgenstern solution) of Γ.

The first condition states that any two imputations of V are incomparable with each other. No domination relation exists between them. This property is known as the *inner stability* of V. The second condition is called the *external stability* of V. This property means that any imputation w not in V is dominated by at least one imputation z of V. That is to say, to any imputation not in V, at least one coalition S will raise objection. This coalition S will strive for a better distribution scheme $z \in V$ such that $z \succ_S w$.

Many cooperative games possess an infinite number of stable sets. A relation between the core and stable sets of a cooperative game is shown in the following theorem.

Theorem 4.8. *Let* $\Gamma \equiv [I, v]$ *be an n-person cooperative game. If the core* $C(\Gamma) \neq \varnothing$, *and* V *is a* vN–M *solution of* Γ, *then* $C(\Gamma) \subseteq V$.

Proof. Assume that $x \in C(\Gamma)$. Then x is undominated by any imputation. If $x \notin V$, then there exists $y \in V$ such that $y \succ x$. This contradicts the supposition that x is not dominated. Therefore, $x \in V$.   □

An inessential cooperative game has only one imputation. Hence, in what follows we deal with essential cooperative games. As was done in the discussion of the core in the preceding section, we consider first constant-sum three-person cooperative games and then proceed to non-constant-sum games.

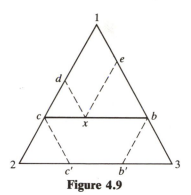

Figure 4.9

Let $\Gamma \equiv [I, v]$ be the essential constant-sum three-person cooperative game in $(0, 1)$ normalization. Then

$$v(\{i\}) = 0, \qquad i = 1, 2, 3,$$
$$v(\{1, 2\}) = v(\{1, 3\}) = v(\{2, 3\}) = 1,$$
$$v(\{1, 2, 3\}) = 1.$$

It is easily seen that in the barycentric coordinates in an equilateral triangle, the imputations represented by the points on a line segment parallel to a side of the triangle, say the line segment bc parallel to the side 23, possesses the following two properties (see Fig. 4.9):

(1) If x, y are any two points on bc, then x does not dominate y and y does not dominate x.

(2) Since

$$\text{Dom}_{\{2,3\}}\, x = xd1e, \quad \text{excluding } xd \text{ and } xe,$$

letting x vary on bc, we obtain

$$\text{Dom}_{\{2,3\}}\, bc = \text{the triangle } bc1, \quad \text{excluding } bc.$$

Similarly,

$$\text{Dom}_{\{1,3\}}\, b = bb'2c, \quad \text{excluding } bb' \text{ and } bc,$$
$$\text{Dom}_{\{1,2\}}\, c = cb3c', \quad \text{excluding } cc' \text{ and } bc.$$

Hence

$$\text{Dom}\, bc = X \backslash bc,$$

i.e. every point of the triangle 123 not lying on bc must be dominated by some point on bc.

It should be noted that in order to have

$$(\text{Dom}_{\{1,3\}}\, b) \cup (\text{Dom}_{\{1,2\}}\, c) = b32c \backslash bc,$$

it is necessary and sufficient that the segment bc be situated below the line joining the mid-points of the sides 12 and 13. That is to say, if the equation of bc is $x_1 = k$, then $0 \leqslant k < \frac{1}{2}$. Under this condition, the line segment bc parallel to the sides 23 of the triangle discussed above is a stable set of the game Γ by definition, i.e. a vN–M solution of Γ.

Since k is an arbitrary non-negative real number less than $\frac{1}{2}$, there are an infinite number of such line segments parallel to the sides 23, and hence an infinite number of vN–M solutions of this kind.

Similarly, the line segments parallel respectively to the sides 12 and 13 of the triangle and situated on the left and right of the lines joining the mid-points of the other two sides are also vN–M solutions of Γ. There are also an infinite number of vN–M solutions for each of these two classes.

The property (1) above is the inner stability of a vN–M solution; property (2) is the external stability.

In addition to the above line segments parallel to a side of the imputation triangle, the constant-sum three-person cooperative game has another stable set $\{a, b, c\}$ consisting of the three mid-points of the sides of the imputation triangle; see Fig. 4.10. The fact that $\{a, b, c\}$ is really a vN–M solution of Γ can be verified by definition.

(1) *Inner stability.* No domination exists between any two of the three imputations a, b, c.

(2) *External stability.* We have

$$\text{Dom}\{a, b, c\} = (\text{Dom}\ a) \cup (\text{Dom}\ b) \cup (\text{Dom}\ c),$$

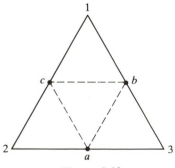

Figure 4.10

where

$$\text{Dom } a = (ab1c)\backslash\{ab,\ ca\},$$
$$\text{Dom } b = (bc2a)\backslash\{bc,\ ab\},$$
$$\text{Dom } c = (ca3b)\backslash\{ca,\ bc\}.$$

But

$$\text{Dom } a \supset (bc\backslash\{b,\ c\}),$$
$$\text{Dom } b \supset (ca\backslash\{c,\ a\}),$$
$$\text{Dom } c \supset (ab\backslash\{a,\ b\}).$$

Hence

$$\text{Dom}\{a,\ b,\ c\} = X\backslash\{a,\ b,\ c\}.$$

Thus any point in the imputation triangle except a, b and c is dominated by at least one of the three points a, b, c. Therefore, the three imputations

$$a = (0,\ \tfrac{1}{2},\ \tfrac{1}{2}), \qquad b = (\tfrac{1}{2},\ 0,\ \tfrac{1}{2}), \qquad c = (\tfrac{1}{2},\ \tfrac{1}{2},\ 0)$$

form a vN–M solution of the game Γ. This solution is called the *symmetric* solution of Γ. The solutions of the line segments described previously are called *discriminatory* solutions. The player who receives the constant payoff k $(0 \leqslant k < \tfrac{1}{2})$ is called the discriminated player.

We have analysed the vN–M solutions for an essential constant-sum three-person cooperative game. For the original detailed discussion see von Neumann and Morgenstern (1953, pp. 285–288).

Now consider the essential non-constant-sum three-person cooperative game $\Gamma \equiv [I, v]$. Suppose that the values of the characteristic function v are

$$v(\{i\}) = 0, \qquad i = 1, 2, 3,$$
$$v(\{1,\ 2\}) = c_3,$$
$$v(\{1,\ 3\}) = c_2,$$
$$v(\{2,\ 3\}) = c_1,$$
$$v(\{1,\ 2,\ 3\}) = 1,$$

where c_1, c_2, c_3 are real numbers in the closed interval $[0, 1]$.

By the definition (4.49) of the core, it is clear that the imputation $x = (x_1, x_2, x_3) \in C(\Gamma)$ if and only if

$$v(\{1,\ 2\}) + v(\{1,\ 3\}) + v(\{2,\ 3\}) = c_1 + c_2 + c_3 \leqslant 2. \qquad (4.57)$$

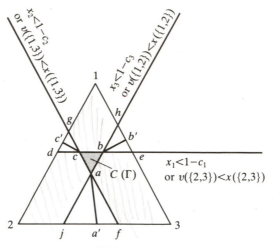

Figure 4.11

Thus the core is empty if and only if

$$v(\{1, 2\}) + v(\{1, 3\}) + v(\{2, 3\}) = c_1 + c_2 + c_3 > 2. \qquad (4.58)$$

This condition was mentioned at the end of Section 4.7.

We shall take up the case $C(\Gamma) \neq \emptyset$ first. In this case (4.57) holds. We noted in Section 4.7 that the core of the game Γ can be a point, a line segment, a triangle, a quadrangle, a pentagon, or a hexagon. Let the core $C(\Gamma)$ of Γ be as shown in Fig. 4.11. The points in $C(\Gamma)$ are the imputations undominated by any other imputations.

Consider the feasibility condition (4.27) of domination of imputations. With respect to the coalition $\{1, 2\}$, in the core $C(\Gamma)$, the condition (4.27) is satisfied only by the points on the line segment ab, where

$$v(\{1, 2\}) = x(\{1, 2\}).$$

The dominion of ab with respect to the coalition $\{1, 2\}$ is

$$\text{Dom}_{\{1,2\}}\, ab = abe3fa, \quad \text{excluding } af,\ be.$$

Similarly,

$$\text{Dom}_{\{1,3\}}\, ca = caj2dc, \quad \text{excluding } cd,\ aj,$$

$$\text{Dom}_{\{2,3\}}\, bc = bcg1hb, \quad \text{excluding } bh,\ cg.$$

Hence the dominion of the boundary ab, bc, ca of $C(\Gamma)$ consists of all the interior points of the three pentagons plus portions of the sides of the imputation triangle; see Fig. 4.11. On the other hand, all points of the three closed triangles *afj*, *bhe*, *cdg* are undominated by any point of the boundary of $C(\Gamma)$. The interior points of $C(\Gamma)$ do not satisfy the effectiveness condition of domination with respect to any two-person coalition.

Let us now examine the triangle *afj*. It is a set of imputations undominated by any point of $C(\Gamma)$. Conversely, any point of $C(\Gamma)$ is undominated by any point of the triangle. We draw a line *aa'* connecting the vertex *a* and an arbitrary point *a'* on the opposite side *fj* of the triangle *afj*. Then it is not difficult to see that any two points on the line *aa'* do not dominate each other. In fact, with respect to the coalition $\{2,3\}$, the imputations on *aa'* do not satisfy the effectiveness condition of domination of imputations. At the same time, with respect to the coalition $\{1,2\}$, although the effectiveness condition is satisfied, any two points

$$x = (x_1, x_2, x_3) \qquad \text{and} \qquad x' = (x'_1, x'_2, x'_3)$$

on *aa'* must have either

$$x_1 > x'_1, \qquad x_2 \leqslant x'_2,$$

or

$$x_1 < x'_1, \qquad x_2 \geqslant x'_2.$$

With respect to the coalition $\{1,3\}$, the situation is similar. Let *x* be any point on *aa'*; see Fig. 4.12. Then

$$\text{Dom}_{\{1,2\}} x = xq3s, \qquad \text{Dom}_{\{1,3\}} x = xr2p.$$

Letting *x* vary along *aa'* from *a* to *a'*, we see that Dom *aa'* contains the whole triangle *afj* except the line segment *aa'*.

Analogously, we draw the lines *bb'*, *cc'* joining respectively the other

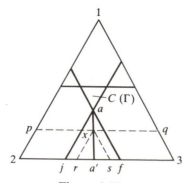

Figure 4.12

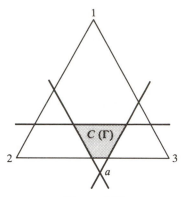

Figure 4.13

two vertices b, c of the triangle abc and two arbitrary points b', c' on the opposite sides he, dg of the small triangles bhe, cdg; see Fig. 4.11. Then Dom bb' contains the whole triangle bhe except the line segment bb', Dom cc' contains the whole triangle cdg except cc'.

Summing up, we obtain a vN–M solution V of the game Γ:

$$V = C(\Gamma) \cup aa' \cup bb' \cup cc'.$$

That is to say, if a non-constant-sum three-person cooperative game Γ has a core $C(\Gamma) \neq \varnothing$, then by Theorem 4.8, $C(\Gamma) \subseteq V$. Now if the graph of $C(\Gamma)$ is as shown in Fig. 4.11, then by adding three line segments aa', bb', cc' to the $C(\Gamma)$ we obtain a vN–M solution V of Γ.

In the case that the core of Γ is not a triangle, e.g. the point of intersection a of

$$x_2 = 1 - c_2 \quad \text{and} \quad x_3 = 1 - c_3$$

is situated outside of the imputation triangle X, as in Fig. 4.13, then $C(\Gamma)$ is a quadrangle, and the undominated triangle afj disappears. Only two line segments are required to add to $C(\Gamma)$ in order to obtain a vN–M solution.

The other cases are similar and will not be enumerated here.

We now turn to the discussion of vN–M solutions of an essential non-constant-sum three-person cooperative game with empty core. Suppose as before that the values of the characteristic function v are

$$v(\{i\}) = 0, \qquad i = 1, 2, 3,$$
$$v(1, 2\}) = c_3,$$
$$v(\{1, 3\}) = c_2,$$
$$v(\{2\,3\}) = c_1,$$
$$v(\{1, 2, 3\}) = 1.$$

It has been pointed out that $C(\Gamma) = \emptyset$ if and only if the inequality (4.58)

$$c_1 + c_2 + c_3 > 2$$

holds. In this case, the lines

$$x_1 = 1 - c_1 \qquad \text{(i.e. } v(\{2, 3\}) = x(\{2, 3\})),$$
$$x_2 = 1 - c_2 \qquad \text{(i.e. } v(\{1, 3\}) = x(\{1, 3\})),$$
$$x_3 = 1 - c_3 \qquad \text{(i.e. } v(\{1, 2\}) = x(\{1, 2\}))$$

are situated in the imputation triangle X as shown in Fig. 4.14.

No point of the small triangle abc is dominated by any point outisde abc. Moreover, every point of abc satisfies the effectiveness condition (4.27) of domination of imputations with respect to each of the three coalitions $\{1, 2\}$, $\{1, 3\}$, $\{2, 3\}$. Therefore, as in the discussion of constant-sum three-person cooperative games, we can consider a line segment de parallel to the side bc of the small triangle abc.

The line segment de has the following properties:

(1) Any two imputations on de do not dominate each other.

(2) The dominions of de with respect to the three two-player coalitions are (see Fig. 4.15):

$$\text{Dom}_{\{1,2\}}\, de = dh3j,$$
$$\text{Dom}_{\{1,3\}}\, de = em2g,$$
$$\text{Dom}_{\{2,3\}}\, de = del1kd.$$

Hence, the region undominated by de is

$$fjm \cup elh \cup dgk.$$

We add to the three triangles fjm, elh, dgk, the line segments ff', ee',

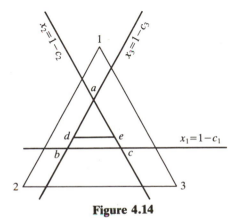

Figure 4.14

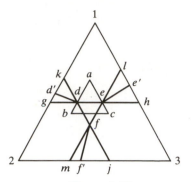

Figure 4.15

dd', respectively; then these line segments together with de constitute a vN–M solution V:

$$V = de \cup ff' \cup ee' \cup dd'.$$

Finally, let us consider the three mid-points d, e, f of the three sides of the triangle abc. These points do not dominate each other, and their dominions are (see Fig. 4.16):

$$\text{Dom}_{\{2,3\}}\, d = dk1l,$$

$$\text{Dom}_{\{1,3\}}\, e = em2g,$$

$$\text{Dom}_{\{1,2\}}\, f = fh3j.$$

In the triangles djm, elh, fgk undominated by $\{d, e, f\}$, we add the line segments dd', ee', ff', respectively; then these line segments including

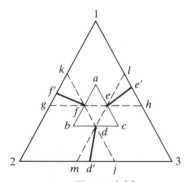

Figure 4.16

the end points constitute another vN–M solution V for the game:

$$V = dd' \cup ee' \cup ff'.$$

For the original detailed discussion of the vN–M solutions for an essential non-constant-sum three-person cooperative game see von Neumann and Morgenstern (1953, pp. 406–415).

The notion of stable sets is obviously an unsatisfactory concept of solution. Moreover, Lucas (1969) gave the following example of a 10-person cooperative game for which there does not exist a stable set:

$$\Gamma \equiv [I, v], \qquad I = \{1, \ldots, 10\};$$

$$v(I) = 5, \qquad v(\{1, 3, 5, 7, 9\}) = 4,$$

$$v(\{1, 2\}) = v(\{3, 4\}) = v(\{5, 6\}) = v(\{7, 8\}) = v(\{9, 10\}) = 1,$$

$$v(\{3, 5, 7, 9\}) = v(\{1, 5, 7, 9\}) = v(\{1, 3, 7, 9\}) = 3,$$

$$v(\{3, 5, 7\}) = v(\{1, 5, 7\}) = v(\{1, 3, 7\}) = 2,$$

$$v(\{3, 5, 9\}) = v(\{1, 5, 9\}) = v(\{1, 3, 9\}) = 2,$$

$$v(\{1, 4, 7, 9\}) = v(\{3, 6, 7, 9\}) = v(\{5, 2, 7, 9\}) = 2,$$

$$v(S) = 0 \qquad \text{for all other } S \subset I.$$

The core of this game is non-empty.

Lucas and Rabie (1982) described another game with 14 players which has no stable set and for which the core is empty.

These games are not superadditive. However, they can be transformed into superadditive games without altering their cores or stable sets (see Lucas, 1969; Lucas and Rabie, 1982).

4.9   Pre-imputations and strong ε-cores

We introduce some further notions which will be used in the following sections. First we give a new definition of an n-person cooperative game.

Definition 4.9. We call $\Gamma \equiv [I, v]$ an n-person cooperative game, where $I = \{1, \ldots, n\}$; v is a real-valued function defined on the set of all subsets of I, satisfying

$$v(\emptyset) = 0, \tag{4.59}$$

$$v(I) \geqslant \sum_{i=1}^{n} v(\{i\}). \tag{4.60}$$

v is called the characteristic function of the game Γ.

This definition is exactly the same as Definition 4.1. However, here we no longer treat $v(S)$ as the value of the matrix game between the two players S and $I \backslash S$. That is to say, when some players of I form a coalition S, we do not suppose the remaining players will form a coalition $I \backslash S$ to act against the interests of S. In this case, we can no longer use eqn (4.3) as a definition of $v(S)$.

The characteristic function v in Definition 4.9 is just a set function which is required to satisfy the two conditions (4.59), (4.60) and no more. Of course, the interesting cases in (4.60) are those for which the inequality is strict.

This definition will no longer guarantee the superadditivity (4.4) of the characteristic function. In this sense, the n-person cooperative game is more generalized than those considered in the previous sections. Of course, the side payments condition is still valid; cf. Section 4.1.

In Section 4.3 an imputation of an n-person cooperative game is defined to be an n-dimensional vector

$$x = (x_1, \ldots, x_n) \tag{4.8}$$

satisfying the condition of individual rationality

$$x_i \geqslant v(\{i\}), \qquad i = 1, \ldots, n \tag{4.9}$$

and the condition of group rationality or Pareto optimality

$$x(I) = v(I). \tag{4.10}$$

Also, the set of all imputations is denoted by X.

We now make an extension of the concept of an imputation. We shall call the n-dimensional vectors satisfying $x(I) = v(I)$ without necessarily being elements of X *pre-imputations* of Γ. In other words, a pre-imputation is a vector (4.8) satisfying (4.10) but not necessarily (4.9). Denote the set of all pre-imputations by X^*.

Definition 4.10. Let $\Gamma \equiv [I, v]$ be an n-person cooperative game. For every pre-imputation $x \in X^*$ and every coalition $S \subseteq I$, define

$$e(S, x) = v(S) - x(S). \tag{4.61}$$

$e(S, x)$ is called the *excess* of S at x.

The excess $e(S, x)$ represents the difference between $v(S)$ and the sum of payoffs that the distribution scheme x offers to the members of S, if this coalition is formed. If $e(S, x)$ is positive, it means that the total proceeds of the members of S exceed the share that x offers them; if

$e(S, x)$ is negative, it means that there is a loss to the coalition S with respect to x.

Using the notation of an excess, (4.9) can be written in the form

$$e(\{i\}, x) \leq 0, \qquad i = 1, \ldots, n; \tag{4.62}$$

and (4.10) can be written in the form

$$e(I, x) = 0. \tag{4.63}$$

The definition of the core $C(\Gamma)$ of an n-person cooperative game Γ in Section 4.7 is

$$C(\Gamma) = \{x : x \in X; v(S) - x(S) \leq 0, S \subseteq I\}. \tag{4.49}$$

Expressed in terms of excess, it becomes

$$C(\Gamma) = \{x : x \in X; e(S, x) \leq 0, S \subseteq I\}. \tag{4.64}$$

This definition is equivalent to the following:

$$C(\Gamma) = \{x : x \in X^*; e(S, x) \leq 0, S \subseteq I\}, \tag{4.65}$$

where X^* is the set of all pre-imputations of Γ.

Definition 4.11. Let $\Gamma \equiv [I, v]$ be an n-person cooperative game, and let ε be a real number. The set of pre-imputations

$$C_\varepsilon(\Gamma) = \{x : x \in X^*; e(S, x) \leq \varepsilon, S \subset I, S \neq \varnothing, I\} \tag{4.66}$$

is called the *strong ε-core* or simply ε-core of Γ.

The strong ε-core consists of all those pre-imputations x at which the excesses of all coalitions $S \neq \varnothing, I$ do not exceed ε.

Obviously, when $\varepsilon = 0$, $C_\varepsilon(\Gamma) = C_0(\Gamma) = C(\Gamma)$. The core $C(\Gamma)$ of Γ is the strong 0-core of Γ. Of course, ε can be a negative number. If ε is sufficiently small, we have $C_\varepsilon(\Gamma) = \varnothing$; if ε is large enough, $C_\varepsilon(\Gamma) \neq \varnothing$. Also, if $\varepsilon_1 < \varepsilon_2$, then $C_{\varepsilon_1}(\Gamma) \subseteq C_{\varepsilon_2}(\Gamma)$.

Definition 4.12. Let $\Gamma \equiv [I, v]$ be an n-person cooperative game. If ε_0 is the smallest ε such that $C_\varepsilon(\Gamma) \neq \varnothing$, then $C_{\varepsilon_0}(\Gamma)$ is called the *least core* of Γ and is denoted by $LC(\Gamma)$.

The least core $LC(\Gamma) = C_{\varepsilon_0}(\Gamma)$ of Γ is the intersection of all non-empty strong ε-cores of Γ.

We illustrate the relations between the core, strong ε-cores and the least core by two examples of three-person cooperative games.

To simplify the expressions, we shall use the following notation:

$$v(1) = v(\{1\}), \qquad x(1) = x_1 = x(\{1\}),$$
$$e(1) = e(\{1\}, x), \qquad v(13) = v(\{1, 3\}),$$
$$x(23) = x(\{2, 3\}), \qquad e(23) = e(\{2, 3\}, x),$$

etc.

Example 4.9. The characteristic function v of the three-person cooperative game Γ of Example 4.7 has the values:

$$v(i) = 0, \qquad i = 1, 2, 3,$$
$$v(12) = \tfrac{1}{3}, \qquad v(13) = \tfrac{1}{6}, \qquad v(23) = \tfrac{5}{6},$$
$$v(123) = 1.$$

Figure 4.17 shows the core $C(\Gamma) = C_0(\Gamma)$ and strong ε-cores for $\varepsilon = \tfrac{1}{6}$
and $\varepsilon = \tfrac{2}{6}$. $C(\Gamma)$ is a quadrangle, $C_{\frac{1}{6}}(\Gamma)$ is a pentagon, $C_{\frac{2}{6}}(\Gamma)$ is a hexagon.
The least core is $C_{-\frac{1}{12}}(\Gamma)$. It is a line segment parallel to the side 23 of the
imputation triangle X:

$$C_{-\frac{1}{12}}(\Gamma) = \{x : x_1 = \tfrac{1}{12}, x_2 \leqslant \tfrac{9}{12}, x_3 \leqslant \tfrac{7}{12}\}.$$

The core is already known to be

$$C(\Gamma) = \{x : x_1 \leqslant \tfrac{1}{6}, x_2 \leqslant \tfrac{5}{6}, x_3 \leqslant \tfrac{4}{6}\}.$$

In the next example the core of the game is empty.

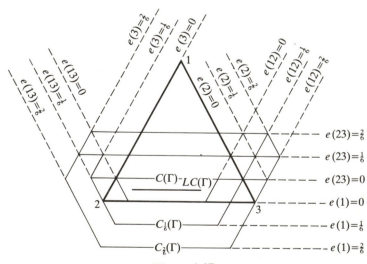

Figure 4.17

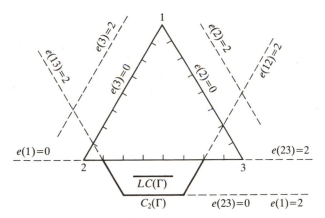

Figure 4.18

Example 4.10. The characteristic function v of a three-person cooperative game Γ has the values:

$$v(i) = 0, \qquad i = 1, 2, 3,$$
$$v(12) = 4, \qquad v(13) = 3, \qquad v(23) = 10,$$
$$v(123) = 8.$$

The characteristic function v is not superadditive. This game has an
empty core. Figure 4.18 shows the strong 2-core $C_2(\Gamma)$ which is a
quadrangle region bounded by the lines $e(1) = 2$, $e(2) = 2$, $e(3) = 2$,
$e(12) = 2$, $e(13) = 2$ and $e(23) = 2$. The least core is

$$LC(\Gamma) = C_1(\Gamma) = \{x: x_1 = -1, x_2 \leqslant 6, x_3 \leqslant 5\}.$$

4.10   The kernel of a cooperative game

This section is concerned with the concept of the kernel and its geometric
properties of an n-person cooperative game.

Let $\Gamma \equiv [I, v]$ be an n-person cooperative game. As was mentioned in
the preceding section, the characteristic function v is not necessarily
superadditive. Denote by T_{ij} the set of all coalitions containing i but not j,
i.e.

$$T_{ij} = \{S: S \subset I, i \in S, j \notin S\}. \tag{4.67}$$

For example, if $I = \{1, 2, 3, 4\}$,

$$T_{42} = \{\{4\}, \{4, 1\}, \{4, 3\}, \{4, 1, 3\}\}.$$

For every pre-imputation $x \in X^*$, define

$$s_{ij}(x) = \max_{S \in T_{ij}} e(S, x). \tag{4.68}$$

It is called the *maximum surplus* of i over j at x. If

$$s_{ij}(x) > s_{ji}(x), \tag{4.69}$$

we say that i *outweighs* j at x. If i does not outweigh j and j does not outweigh i at x, i.e. if

$$s_{ij}(x) = s_{ji}(x), \tag{4.70}$$

we say that i and j are *in equilibrium* at x. These concepts are relative to the set X^* of pre-imputations.

With regard to the imputation set X, the concept of maximum surplus is the same. For the concepts of outweighing and equilibrium, certain modifications are necessary. For any imputation $x \in X$, if

$$s_{ij}(x) > s_{ji}(x) \tag{4.71}$$

and

$$x_j > v(j), \tag{4.72}$$

then i outweighs j at x. If i does not outweigh j, nor does j outweigh i, then i and j are in equilibrium at x.

Hence, i and j are in equilibrium at $x \in X$ if and only if

$$[s_{ij}(x) - s_{ji}(x)][x_j - v(j)] \leqslant 0, \tag{4.73}$$

$$[s_{ji}(x) - s_{ij}(x)][x_i - v(i)] \leqslant 0. \tag{4.74}$$

Definition 4.13. The *pre-kernel* of an n-person cooperative game $\Gamma \equiv [I, v]$ is the set $K^*(\Gamma)$ of pre-imputations $x \in X^*$ at which every two players i, j are in equilibrium with respect to X^*.

This definition implies that a pre-imputation $x \in K^*(\Gamma)$ if and if (4.70) holds for every pair of players i, j.

Definition 4.14. The *kernel* of an n-person cooperative game $\Gamma \equiv [I, v]$ is the set $K(\Gamma)$ of imputations $x \in X$ at which every two players i, j are in equilibrium with respect to X.

Thus, an imputation $x \in X(\Gamma)$ if and only if (4.73), (4.74) hold simultaneously for every pair of players i, j. This is equivalent to

$$s_{ij}(x) = s_{ji}(x) \tag{4.75}$$

or

$$s_{ij}(x) > s_{ji}(x), \qquad x_j = v(j) \tag{4.76}$$

or

$$s_{ji}(x) > s_{ij}(x), \qquad x_i = v(i). \tag{4.77}$$

Comparing these two definitions, it is obvious that the pre-kernel is much easier to compute than the kernel.

It has been proved that the pre-kernel and kernel are always non-empty. Some of the properties of the pre-kernel and kernel have been explored. Readers are referred to Aumann *et al.* (1965), Davis and Maschler (1965), Maschler and Peleg (1966), Maschler *et al.* (1972, 1979), Peleg (1966).

For large n, the computation of the kernel of a cooperative game is not an easy task. We illustrate this by two examples with $n = 3$.

Example 4.11. Consider the three-person cooperative game Γ of Example 4.9. The characteristic function v has the values:

$$v(i) = 0, \qquad i = 1, 2, 3,$$
$$v(12) = \tfrac{1}{3}, \qquad v(13) = \tfrac{1}{6}, \qquad v(23) = \tfrac{5}{6},$$
$$v(123) = 1.$$

To find the pre-kernel $K^*(\Gamma)$ and kernel $K(\Gamma)$, we draw the graph of the equation (4.75) for every pair of i, j. For example, in order to have a graph of

$$s_{13}(x) = s_{31}(x), \tag{4.78}$$

we divide the space X^* of pre-imputations into three regions.

In Fig. 4.19, for the region to the left of the line AB, we have

$$s_{13} = e(1), \qquad s_{31} = e(3).$$

Hence the graph of $s_{13} = s_{31}$ is the bisector of the angle formed by the straight lines $e(1) = 0$ and $e(3) = 0$. Next, in the region between the lines AB and CD,

$$s_{13} = e(1), \qquad s_{31} = e(32).$$

Hence the graph of $s_{13} = s_{31}$ is the bisector of the region between $e(1) = 0$ and $e(32) = 0$, i.e. the line segment equidistant from $e(1) = 0$ and $e(32) = 0$. Similarly, in the region to the right of the line CD, the graph of $s_{13} = s_{31}$ is the bisector of the angle formed by the lines $e(12) = 0$ and $e(32) = 0$. Thus we obtain the complete graph of $s_{13}(x) = s_{31}(x)$, represented in the figure by the heavy black line consisting of three line segments.

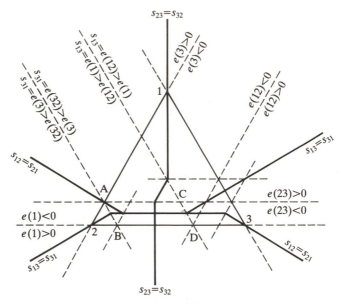

Figure 4.19

In an analogous way we obtain the graphs of the equations

$$s_{12}(x) = s_{21}(x) \tag{4.79}$$

and

$$s_{23}(x) = s_{32}(x). \tag{4.80}$$

The three black lines (4.78), (4.79), (4.80) intersect at the same point. That is to say, at this point of intersection eqn (4.70) or (4.75) holds for every pair of i, j. By Definitions 4.13 and 4.14, every two players i, j are in equilibrium at this point of intersection with respect to X^* and X. Therefore, this point is at the same time the pre-kernel $K^*(\Gamma)$ and the kernel $K(\Gamma)$ of the game Γ. We have

$$K^*(\Gamma) = K(\Gamma) = \{(\tfrac{1}{12}, \tfrac{13}{24}, \tfrac{9}{24})\}.$$

The point is the mid-point of the least core $LC(\Gamma) = C_{-\frac{1}{12}}(\Gamma)$. It is also the centre of the core $C(\Gamma)$; cf. Example 4.9.

The following is an example in which the characteristic function v of the game is not superadditive; it is the game of Example 4.10.

Example 4.12. The characteristic function v of a three-person cooperative game Γ has the values:

$$v(i) = 0, \qquad i = 1, 2, 3,$$
$$v(12) = 4, \qquad v(13) = 3, \qquad v(23) = 10,$$
$$v(123) = 8.$$

We have seen in Example 4.10 that the core $C(\Gamma)$ of this game is an
empty set. Starting from $\varepsilon = 0$, we increase ε gradually and consider the
strong ε-core. At $\varepsilon = 1$, the strong ε-core comes into view, situated
outside the imputation triangle X. This is the least core $LC(\Gamma) = C_1(\Gamma)$;
see Fig. 4.20. If we continue to increase ε, the strong ε-core touches X at
$\varepsilon = 2$.

Using the same method as in Example 4.11, we draw the graphs of

$$s_{12}(x) = s_{21}(x), \tag{4.81}$$

$$s_{13}(x) = s_{31}(x), \tag{4.82}$$

$$x_{23}(x) = s_{32}(x). \tag{4.83}$$

The three black lines intersect at a single point which forms the
pre-kernel $K^*(\Gamma)$ of the game Γ.

At the point which represents the kernel $K(\Gamma)$, only one equation

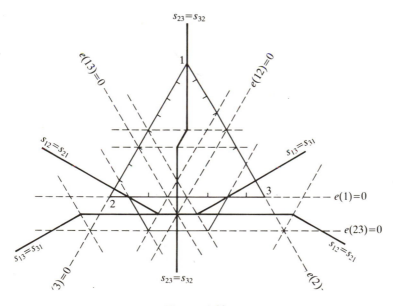

Figure 4.20

(4.83) holds, i.e.

$$s_{23}(x) = s_{32}(x);$$

the other two are replaced by inequalities:

$$s_{21}(x) > s_{12}(x), \qquad x_1 = 0 = v(1),$$
$$s_{31}(x) > s_{13}(x), \qquad x_1 = 0.$$

These two inequalities are of the form (4.76) or (4.77).

By simple calculation, the pre-kernel is found to be

$$K^*(\Gamma) = \{(-1, 5, 4)\}.$$

As in the case of Example 4.11, this point is also the mid-point of the least core $LC(\Gamma) = C_1(\Gamma)$. The kernel of the game is

$$K(\Gamma) = \{0, 4.5, 3.5)\}.$$

The concepts of the pre-kernel and kernel introduced above are defined for the grand coalition I without imposing any condition on the formation of coalitions. We now describe briefly the concept of the kernel for a fixed coalition structure.

Let $\mathcal{B}$ be a *partition* of I, i.e.

$$\mathcal{B} = \{B_1, \ldots, B_p\}, \tag{4.84}$$

where

$$B_j \cap B_k = \varnothing, \qquad j \neq k,$$

$$\bigcup_{k=1}^{p} B_k = I.$$

$\mathcal{B}$ is called a *coalition structure*.

An *individually rational payoff configuration* is a pair

$$(x; \mathcal{B}) = (x_1, \ldots, x_n; B_1, \ldots, B_p), \tag{4.85}$$

where $\mathcal{B}$ is a coalition structure and $x = (x_1, \ldots, x_n)$ is a *payoff vector* satisfying

$$x_i \geq v(i), \qquad i = 1, \ldots, n, \tag{4.86}$$

$$x(B_k) = v(B_k), \qquad k = 1, \ldots, p. \tag{4.87}$$

Relative to a fixed coalition structure $\mathcal{B}$, for a payoff vector x satisfying (4.86) and (4.87), the concepts of excess, maximum surplus, outweighing, and equilibrium are the same as above, except that all imputations are replaced by payoff vectors, and that i and j must be members of the same $B_k \in \mathcal{B}$ whenever the concepts of outweighing and equilibrium are involved. We redefine the notion of the kernel as follows.

Definition 4.15. Let $\mathscr{B}$ be a coalition structure of an n-person cooperative game $\Gamma \equiv [I, v]$. The kernel $K(\Gamma)$ of Γ with respect to $\mathscr{B}$ is the set of all those individually rational payoff configurations $(x; \mathscr{B})$ for which every two players belonging to the same $B_k \in \mathscr{B}$ are, in equilibrium.

If $\mathscr{B} = \{I\}$, the kernel of the game with respect to the grand coalition is the kernel in Definition 4.14.

Example 4.13. Find the kernels of the game of Example 4.12 with respect to the coalition structures

$$\{\{2, 3\}, \{1\}\}, \qquad \{\{1, 3\}, \{2\}\}, \qquad \{\{1, 2\}, \{3\}\}.$$

The characteristic function v has the values

$$v(i) = 0, \qquad i = 1, 2, 3,$$
$$v(12) = 4, \qquad v(13) = 3; \qquad v(23) = 10,$$
$$v(123) = 8.$$

Consider first the coalition structure $\mathscr{B} = \{\{2, 3\}, \{1\}\}$. By (4.87), we have

$$x(1) = x_1 = v(1) = 0, \qquad x(23) = x_2 + x_3 = v(23) = 10.$$

The kernel of the game with respect to the coalition structure $\{\{2, 3\}, \{1\}\}$ is seen to consist of the single point

$$(0, \tfrac{11}{2}, \tfrac{9}{2}).$$

The kernels of the game with respect to the coalition structures $\{\{1, 3\}, \{2\}\}$ and $\{\{1, 2\}, \{3\}\}$ are respectively $\{(0, 0, 3)$ and $\{(0, 4, 0)\}$.

4.11   The nucleolus of a cooperative game

The *nucleolus* is an important solution concept of the n-person cooperative game. Like the kernel, the nucleolus is also a distribution scheme under which the amount $v(I) = x(I)$ is to be divided among the n players.

Let $\Gamma \equiv [I, v]$ be an n-person cooperative game, and suppose the characteristic function v satisfies the conditions

$$v(\varnothing) = 0, \qquad v(I) \geqslant \sum_{i=1}^{n} v(i).$$

For each coalition of players $S \subseteq I$, consider the excess of S at the imputation $x \in X$:

$$e(S, x) = v(S) - x(S).$$

This number reflects the 'attitude' of the coalition S to the imputation x. The larger the number $e(S, x)$ is, the less will be the x acceptable to S. For a fixed x, the coalition which objects most strongly to it is the one with the greatest excess, i.e. the coalition S_0 for which

$$e(S_0, x) = \max_{S \subseteq I} e(S, x).$$

For every $x \in X$, we can enumerate the excesses $e(S, x)$ of the 2^n subsets $S \subseteq I$ at x. We form a 2^n-dimensional vector $\theta(x)$ with these 2^n excesses $e(S, x)$ as its components and arranged in non-increasing order, i.e.

$$\theta(x) = (\theta_1(x), \ldots, \theta_{2^n}(x)), \qquad \theta_i(x) = e(S, x), \qquad S \subseteq I, \quad (4.88)$$

where

$$\theta_1(x) \geqslant \theta_2(x) \geqslant \cdots \geqslant \theta_{2^n}(x). \tag{4.89}$$

That is to say,

$$\theta_i(x) \geqslant \theta_j(x)$$

for all i, j, $1 \leqslant i \leqslant j \leqslant 2^n$.

For different $x, y \in X$, define the *lexicographical ordering* of θ as follows. If there is an index k_0 such that

$$\theta_k(x) = \theta_k(y), \qquad k < k_0, \tag{4.90}$$

$$\theta_{k_0}(x) < \theta_{k_0}(y), \tag{4.91}$$

we say that $\theta(x)$ is lexicographically smaller than $\theta(y)$. This relation is denoted by

$$\theta(x) < \theta(y). \tag{4.92}$$

'Not $\theta(y) < \theta(x)$' is denoted by

$$\theta(x) \lesssim \theta(y). \tag{4.93}$$

Relations (4.90) and (4.91) signify that the first $k_0 - 1$ components of $\theta(x)$ are equal to the corresponding components of $\theta(y)$, and the k_0th component of $\theta(x)$ is smaller than that of $\theta(y)$.

Definition 4.16. The nucleolus of an n-person cooperative game $\Gamma \equiv [I, v]$ is the set $N(\Gamma)$ of imputations $x \in X$ such that $\theta(x)$ is minimal in the lexicographical ordering. That is,

$$N(\Gamma) = \{x : x \in X; \theta(x) \lesssim \theta(y) \text{ for all } y \in X\}. \tag{4.94}$$

This concept was first introduced by D. Schmeidler. In Schmeidler (1969) the existence and uniqueness of the nucleolus of a cooperative game are proved, and some of the properties of the nucleolus are studied. The following are the main results.

Theorem 4.9. *Let $\Gamma \equiv [I, v]$ be an n-person cooperative game. Then the nucleolus $N(\Gamma)$ of the game is non-empty.*

Proof. We first prove that

$$\theta_i(x), \qquad i = 1, \ldots, 2^n \tag{4.95}$$

are continuous functions of x. For this purpose, we write $\theta_i(x)$ in the following form:

$$\theta_i(x) = \max\{\min\{e(S, x): S \in \mathscr{A}\}: \mathscr{A} \subseteq \mathscr{P}(I), |\mathscr{A}| = i\}, \tag{4.96}$$

where $\mathscr{P}(I)$ is the class of all subsets of I (i.e. the power set of I), $\mathscr{A}$ is a subclass of $\mathscr{P}(I)$, and $|\mathscr{A}|$ is the number of elements (i.e. coalitions) of $\mathscr{A}$.

This definition is equivalent to the original definition of $\theta_i(x)$. In fact, if we replace i by $i + 1$ in (4.96), then

$$\theta_{i+1}(x) = \max\{\min\{e(S, x): S \in \mathscr{A}\}: \mathscr{A} \subseteq \mathscr{P}(I), |\mathscr{A}| = i + 1\}. \tag{4.97}$$

Compare (4.96) and (4.97). In (4.96),

$$\{e(S, x): S \in \mathscr{A}\}$$

has i elements. For each of such

$$\{e(S, x): S \in \mathscr{A}\},$$

there is some $\{e(S, x): S \in \mathscr{A}\}$ containing $i + 1$ elements in (4.97) such that the former is a subset of the latter. Hence, every minimum in (4.96) is not less than some minimum in (4.97). Conversely, every minimum in (4.97) is not greater than some minimum in (4.96). Taking the maximum we have

$$\theta_i(x) \geqslant \theta_{i+1}(x).$$

This inequality holds for all $i = 1, \ldots, 2^n - 1$. Therefore, the two definitions are equivalent.

It follows from (4.96) that, since $e(S, x)$ is a continuous function of x, and since min, max are operations of continuous functions, the θ_i, $i = 1, \ldots, 2^n$, in (4.95) are continuous functions of x.

Next, let

$$X_1 = \{x: x \in X; \theta_1(x) \leqslant \theta_1(y) \text{ for all } y \in X\},$$
$$X_{i+1} = \{x: x \in X_i; \theta_{i+1}(x) \leqslant \theta_{i+1}(y) \text{ for all } y \in X_i\},$$
$$i = 1, \ldots, 2^n - 1.$$

The compactness of the set X (X is a bounded closed set) and the continuity of $\theta_i(x)$ imply that $X_1, \ldots, X_{2^n}$ are all compact and non-empty. Obviously,

$$X_{2^n} = N(\Gamma). \qquad \qquad \qquad \square$$

Theorem 4.10. *Let $N(\Gamma)$ be the nucleolus of an n-person cooperative game $\Gamma \equiv [I, v]$. Then $|N(\Gamma)| = 1$, i.e. $N(\Gamma)$ consists of one element.*

Proof. The set X of all imputations is a convex set. Hence, if $x \in N(\Gamma)$, $y \in N(\Gamma)$, $x \neq y$, we must have

$$\theta\left(\frac{x+y}{2}\right) \geq \theta(x) = \theta(y). \qquad (4.98)$$

Let $z = (z_1, \ldots, z_r)$. Define the ordering function $\eta = (\eta_1, \ldots, \eta_r)$ as follows:

$$\eta_i(x) = \max\{\min\{z_j : j \in A\} : A \subseteq \{1, \ldots, r\}, |A| = i\}, \qquad i = 1, \ldots, r.$$

An equivalent definition of η is that in (1) and (2) below.

(1) $\eta_1(z) = \max\{z_t : t = 1, \ldots, r\}$.
(2) If we have defined $\eta_1(z) = z_{j_1}, \ldots, \eta_i(z) = z_{j_i}$, then

$$\eta_{i+1}(z) = \max\{z_t : t \neq j_1, \ldots, j_i\},$$

where $i = 1, \ldots, r - 1$.

This function is an ordering function which arranges the components of the vector z in non-increasing order.

We now prove that for any two points u, w of R^r we have

$$\eta(u + w) \leq \eta(u) + \eta(w). \qquad (4.99)$$

Let

$$\eta_t(u) = u_{i_t}, \qquad \qquad t = 1, \ldots, r,$$
$$\eta_t(w) = w_{j_t}, \qquad \qquad t = 1, \ldots, r,$$
$$\eta_t(u + w) = u_{k_t} + w_{k_t}, \qquad t = 1, \ldots, r.$$

Obviously,

$$u_{k_1} \leq u_{i_1}, \qquad w_{k_1} \leq w_{j_1}. \qquad (4.100)$$

Hence

$$\eta_1(u + w) = u_{k_1} + w_{k_1} \leq u_{i_1} + w_{j_1} = \eta_1(u) + \eta_1(w). \qquad (4.101)$$

If the inequality '<' holds in (4.101), then

$$\eta(u + w) < \eta(u) + \eta(w). \qquad (4.102)$$

If equality holds in (4.101), then, by (4.100),

$$u_{k_1} = u_{i_1}, \qquad w_{k_1} = w_{j_1}.$$

Now if

$$u_{k_t} = u_{i_t}, \qquad w_{k_t} = w_{j_t} \tag{4.103}$$

for all $t = 1, \ldots, r$, then

$$\eta_t(u + w) = \eta_t(u) + \eta_t(w), \qquad t = 1, \ldots, r,$$

so that

$$\eta(u + w) = \eta(u) + \eta(w). \tag{4.104}$$

It remains to consider the case in which (4.103) does not hold for some t. We may assume that for some s, $1 \leqslant s < r$, we have

$$u_{k_t} = u_{i_t}, \qquad w_{k_t} = w_{j_t}, \qquad t = 1, \ldots, s, \tag{4.105}$$

while

$$u_{k_{s+1}} \neq u_{i_{s+1}}. \tag{4.106}$$

We obtain

$$u_{k_{s+1}} \leqslant \max\{u_i : i \neq k_1, \ldots, k_s\} = \max\{u_i : i \neq i_1, \ldots, i_s\}. \tag{4.107}$$

By (2) in the definition of the ordering function η,

$$\eta_{s+1}(u) = u_{i_{s+1}} = \max\{u_i : i \neq i_1, \ldots, i_s\}. \tag{4.108}$$

It follows from (4.106), (4.107), (4.108) that

$$u_{k_{s+1}} < u_{i_{s+1}}. \tag{4.109}$$

Similarly, from the second equation in (4.105) and the definition of η it follows that

$$w_{k_{s+1}} \leqslant w_{j_{s+1}}. \tag{4.110}$$

By (4.109), (4.110) and the definition of η,

$$\eta_{s+1}(u + w) = u_{k_{s+1}} + w_{k_{s+1}} < u_{i_{s+1}} + w_{j_{s+1}} = \eta_{s+1}(u) + \eta_{s+1}(w).$$

Hence

$$\eta(u + w) < \eta(u) + \eta(w). \tag{4.111}$$

Relations (4.102), (4.104) together with (4.111) prove the validity of (4.99).

Moreover, from the above proof it is seen that if

$$\eta(u + w) = \eta(u) + \eta(w),$$

then

$$u_{k_t} = u_{i_t}, \qquad w_{k_t} = w_{j_t}, \qquad t = 1, \ldots, r. \tag{4.112}$$

We now return to the vectors

$$\{e(S, x): S \subseteq I\},$$
$$\{e(S, y): S \subseteq I\},$$
$$\{e(S, x) + e(S, y): S \subseteq I\}$$

of the space R^{2^n}. By applying the ordering function η to these vectors, it is easily seen that

$$\eta\{e(S, x): S \subseteq I\} = \theta(x),$$
$$\eta\{e(S, y): S \subseteq I\} = \theta(y),$$
$$\eta\{e(S, x) + e(S, y): S \subseteq I\} = \eta\{2v(S) - x(S) - y(S): S \subseteq I\}$$
$$= \eta\left\{2\left[v(S) - \frac{x+y}{2}(S)\right]: S \subseteq I\right\}$$
$$= 2\theta\left(\frac{x+y}{2}\right).$$

Using (4.99) above, we obtain

$$2\theta\left(\frac{x+y}{2}\right) \leqslant \theta(x) + \theta(y). \tag{4.113}$$

The case

$$2\theta\left(\frac{x+y}{2}\right) < \theta(x) + \theta(y)$$

contradicts (4.98).

Suppose

$$2\theta\left(\frac{x+y}{2}\right) = \theta(x) + \theta(y).$$

Let

$$\theta_t(x) = e(S_{i_t}, x), \qquad t = 1, \ldots, 2^n,$$
$$\theta_t(y) = e(S_{j_t}, y), \qquad t = 1, \ldots, 2^n,$$
$$2\theta_t\left(\frac{x+y}{2}\right) = e(S_{k_t}, x) + e(S_{k_t}, y), \qquad t = 1, \ldots, 2^n.$$

Then (4.112) implies

$$e(S_{k_t}, x) = e(S_{i_t}, x), \qquad e(S_{k_t}, y) = e(S_{j_t}, y)$$

for all $t = 1, \ldots, 2^n$. Using the assumption $\theta(x) = \theta(y)$ we get

$$e(S_{k_t}, x) = e(S_{k_t}, y), \qquad t = 1, \ldots, 2^n,$$

or

$$v(S_{k_t}) - x(S_{k_t}) = v(S_{k_t}) - y(S_{k_t}), \qquad t = 1, \ldots, 2^n,$$

or

$$x(S_{k_t}) = y(S_{k_t}), \qquad t = 1, \ldots, 2^n.$$

That is to say, the equation

$$x(S) = y(S)$$

holds for all $S \subseteq I$. Hence $x = y$, contrary to the assumption $x \neq y$.
    Thus we have proved that if $x \in N(\Gamma)$ and $y \in N(\Gamma)$ then $x = y$. $\square$

    Theorems 4.9 and 4.10 show that, for every cooperative game Γ there exists a unique imputation $x \in N(\Gamma)$. The next theorem states that this one-element nucleolus $N(\Gamma)$ is contained in the kernel of Γ.

Theorem 4.11. *Let the kernel and nucleolus of an n-person cooperative game* $\Gamma \equiv [I, v]$ *be respectively* $K(\Gamma)$ *and* $N(\Gamma)$. *Then* $N(\Gamma) \subseteq K(\Gamma)$.

Proof. Assume that $x \in N(\Gamma)$, $x \notin K(\Gamma)$; then there exist i, j such that i outweighs j at x. Thus there is a sufficiently small $\varepsilon > 0$ satisfying

$$s_{ij}(x) > s_{ji}(x), \qquad x_j - \varepsilon > v(j). \tag{4.114}$$

Let α be another sufficiently small positive number (we shall use it later immediately after (4.125)). Define

$$\delta = \tfrac{1}{2} \min\{\varepsilon, s_{ij}(x) - s_{ji}(x), \alpha\}. \tag{4.115}$$

Consider the imputation

$$x^\delta = x - \delta e^j + \delta e^i, \tag{4.116}$$

where e^j is the unit vector whose jth component is 1, and x^δ is the imputation obtained by transferring an amount of δ from player j to player i at x. By (4.115) and the second inequality in (4.114), we have

$$x_j - \delta > x_j - \varepsilon > v(j).$$

Hence x^δ is still in X.
    We shall proceed to show that

$$\theta(x^\delta) < \theta(x). \tag{4.117}$$

The validity of (4.117) will imply that $x \notin N(\Gamma)$, which is in contradiction to the assumption, and the proof of the theorem will be completed.
    We divide the 2^n subsets S of I into three classes and compare the excesses of the members in each class at x and x^δ.

(1) Consider the S in $T_{ij} = \{S: i \in S, j \notin S\}$. There is at least one of these S, say R, whose excess at x is

$$e(R, x) = s_{ij}(x). \tag{4.118}$$

Each of the other S satisfies

$$e(S, x) \leq s_{ij}(x). \tag{4.119}$$

The excess of R at x^{δ} is

$$
\begin{aligned}
e(R, x^{\delta}) &= v(R) - [x(R) + \delta] \\
&= e(R, x) - \delta = s_{ij}(x) - \delta.
\end{aligned} \tag{4.120}
$$

Each of the other S satisfies

$$e(S, x^{\delta}) = e(S, x) - \delta \leq s_{ij}(x) - \delta. \tag{4.121}$$

(2) The S in $T_{ji} = \{S: j \in S, i \notin S\}$. The excess of each of these S at x is

$$e(S, x) \leq s_{ji}(x) < s_{ij}(x). \tag{4.122}$$

The excess of each of these S at x^{δ} is

$$
\begin{aligned}
e(S, x^{\delta}) &= e(S, x) + \delta \\
&\leq s_{ji}(x) + \delta \leq s_{ij}(x) - \delta.
\end{aligned} \tag{4.123}
$$

(3) For the S in $\{S: i, j \in S \text{ or } i, j \notin S\}$, we have

$$e(S, x^{\delta}) = e(S, x).$$

Suppose there are k coalitions S which satisfy

$$e(S, x^{\delta}) = e(S, x) \geq s_{ij}(x). \tag{4.124}$$

All other S satisfy

$$e(S, x^{\delta}) = e(S, x) < s_{ij}(x). \tag{4.125}$$

We may assume that all these latter S satisfy

$$e(S, x^{\delta}) < s_{ij}(x) - \alpha,$$

where $\alpha > 0$ is a sufficiently small number. Hence, by (4.115),

$$e(S, x^{\delta}) < s_{ij}(x) - \delta. \tag{4.126}$$

In view of (4.118)–(4.126), by definition of the ordering function η or θ, since $x \in N(\Gamma)$, we have

$$\theta_m(x^{\delta}) = \theta_m(x) \geq s_{ij}(x), \qquad m = 1, \ldots, k, \tag{4.127}$$

$$
\begin{aligned}
\theta_{k+1}(x^{\delta}) &= e(R, x^{\delta}) = s_{ij}(x) - \delta \\
&< s_{ij}(x) = e(R, x) = \theta_{k+1}(x).
\end{aligned} \tag{4.128}
$$

It follows from (4.127), (4.128) that

$$\theta(x^\delta) < \theta(x).$$

This is the relation (4.117) to be proved.   □

According to this theorem, if the kernel $K(\Gamma)$ of an n-person cooperative game Γ consists of a single point, then this point is also the point of the nucleolus of Γ. For example, the kernels of the games of Example 4.11 and Example 4.12 in the preceding section are both one-point sets. Therefore in each case the nucleolus is identical with the kernel.

We point out in passing that, if the least core of a game Γ is a line segment in the imputation set, the nucleolus of Γ is not in general the mid-point of the line segment; cf. Maschler *et al.* (1979) for an example with $n = 4$.

The above theorem not only shows that the nucleolus $N(\Gamma)$ is contained in the kernel $K(\Gamma)$ of a cooperative game Γ, it provides a new proof for the existence of $K(\Gamma)$. It is a much simpler proof than any earlier existence proof of $K(\Gamma)$.

The nucleolus of this section is defined for the grand coalition. As in the case of the kernel, the concept of the nucleolus can also be defined with respect to a fixed coalition structure. This generalization will not be pursued here.

In Maschler *et al.* (1979) the 'lexicographic centre' of a game is defined, which amounts to an alternative definition of the nucleolus. It has been shown that the lexicographic centre coincides with the nucleolus.

Definition 4.17. Let $\Gamma \equiv [I, v]$ be an n-person cooperative game. Denote $X^0 = X$ and

$$\Sigma^0 = \{S: S \subset I; S \neq \emptyset, I\}.$$

Construct a sequence $X^0 \supset X^1 \supset \cdots \supset X^\kappa$ of sets of imputations and a sequence $\Sigma^0 \supset \Sigma^1 \supset \cdots \supset \Sigma^\kappa$ of sets of coalitions. For $k = 1, \ldots, \kappa$ define recursively

$$\varepsilon^k = \min_{x \in X^{k-1}} \max_{S \in \Sigma^{k-1}} e(S, x), \qquad (4.129)$$

$$X^k = \left\{ x: x \in X^{k-1}; \max_{S \in \Sigma^{k-1}} e(S, x) = \varepsilon^k \right\}, \qquad (4.130)$$

$$\Sigma_k = \{S: S \in \Sigma^{k-1}; e(S, x) = \varepsilon^k \text{ for all } x \in X^k\}, \qquad (4.131)$$

$$\Sigma^k = \Sigma^{k-1} \backslash \Sigma_k, \qquad (4.132)$$

where κ is the first value of k for which $\Sigma^k = \emptyset$.

The set X^κ is called the *lexicographic centre* of Γ.

Theorem 4.12. *The number κ of Definition 4.17 is finite. For $k = 1, \ldots, \kappa$:*
   (i) *the ε^k are finite;*
   (ii) *the X^k are non-empty, compact and convex;*
   (iii) *$\Sigma_k \neq \emptyset$;*
      and for $k = 1, \ldots, \kappa - 1$:
   (iv) *$\varepsilon^{k+1} < \varepsilon^k$.*

Proof. Since X^0 satisfies (ii)$_0$, (ii)$_0$ implies (i)$_1$, and in general (i)$_k$ implies (ii)$_k$, which in turn implies (i)$_{k+1}$ so long as $\Sigma^k \neq \emptyset$, both (i)$_k$ and (ii)$_k$ are proved by induction up to $k = \kappa$.
   Claim (iii)$_\kappa$ follows from the fact that

$$\Sigma_\kappa = \Sigma^{\kappa-1} \backslash \Sigma^\kappa = \Sigma^{\kappa-1} \backslash \emptyset = \Sigma^{\kappa-1} \neq \emptyset.$$

To prove (iii)$_k$ for $k < \kappa$, let $S_1, \ldots, S_m$ be the members of $\Sigma^k \neq \emptyset$. By (4.131) and (4.132), for each $j = 1, \ldots, m$, there exists an imputation $x^{(j)} \in X^k$ such that $e(S, x^{(j)}) < \varepsilon^k$. Define

$$\bar{x} = \frac{1}{m} \sum_{j=1}^{m} x^{(j)}. \tag{4.133}$$

By convexity, $\bar{x} \in X^k$. Hence there exists a coalition $\bar{S} \in \Sigma^{k-1}$ with

$$e(\bar{S}, \bar{x}) = \varepsilon^k. \tag{4.134}$$

Since $e(\bar{S}, x^{(j)}) \leqslant \varepsilon^k$, $j = 1, \ldots, m$, we have

$$e(\bar{S}, \bar{x}) = v(\bar{S}) - \bar{x}(\bar{S}) = v(\bar{S}) - \frac{1}{m} \sum_{j=1}^{m} x^{(j)}(\bar{S})$$

$$= \frac{1}{m} \sum_{j=1}^{m} e(\bar{S}, x^{(j)}) \leqslant \varepsilon^k. \tag{4.135}$$

Now if $\bar{S} \in \Sigma^k$ (i.e. if $\bar{S}$ be one of the S_j), then the strict inequality

$$e(\bar{S}, \bar{x}) < \varepsilon^k \tag{4.136}$$

must hold. This contradiction proves that

$$\bar{S} \in \Sigma^{k-1} \backslash \Sigma^k = \Sigma_k,$$

i.e. $\Sigma_k \neq \emptyset$, and the proof of (iii) is completed. The finiteness of κ follows from the fact that at each stage a non-empty set Σ_k of coalitions is removed from Σ^{k-1} to obtain Σ^k.
   Finally, to prove (iv), consider the same imputation $\bar{x} \in X^k$ defined in (4.133). As remarked at (4.135), the strict inequality $e(S, \bar{x}) < \varepsilon^k$ holds for all $S \in \Sigma^k$. Therefore, for $k < \kappa$ we have

$$\varepsilon^{k+1} \leqslant \max_{S \in \Sigma^k} e(S, \bar{x}) < \varepsilon^k. \qquad \square$$

Theorem 4.13. *The lexicographic centre of a game* Γ *consists of one element.*

Proof. By Theorem 4.12(ii), the lexicographic centre X^κ of Γ is non-empty. By (4.131), the excess of each S is constant in X^k if $X \in \Sigma_k$, and hence is constant in the lexicographic centre X^κ. In particular, this is true for the single-person coalitions. Consequently the components of the imputations in X^κ are all constant and therefore it is impossible for X^κ to contain more than one imputation. $\qquad\square$

Theorem 4.14. *The nucleolus of an n-person cooperative game* $\Gamma \equiv [I, v]$ *coincides with its lexicographic centre.*

Proof. It suffices to prove that for any k, $1 \leq k \leq \kappa$, if $x \in X^k$ and $y \in X^{k-1} \backslash X^k$, then $\theta(x) < \theta(y)$.

Consider the partititon of the set of all subsets of I into the sets $\Sigma_1, \ldots, \Sigma_{k-1}, \Sigma^{k-1}, \{\varnothing, I\}$. We may ignore $\varnothing$ and I in the lexicographic comparison between $\theta(x)$ and $\theta(y)$, since their excesses are always zero. By (4.131) and Theorem 4.12(iv), for all $S \in \Sigma_h$, $h = 1, \ldots, k-1$, we have

$$e(S, x) = e(S, y) = \varepsilon^h \geq \varepsilon^{k-1}.$$

Moreover, for all $S \in \Sigma^{k-1}$ we have

$$e(S, x) \leq \varepsilon^k < \varepsilon^{k-1}, \qquad e(S, y) \leq \varepsilon^{k-1}$$

by (4.130). However, since y is not in X^k, there exists at least one R in Σ^{k-1} for which

$$\varepsilon^k < e(R, y) \leq \varepsilon^{k-1}.$$

Hence (4.90) and (4.91) are satisfied for the index k_0 corresponding to some such R. Consequently we have $\theta(x) < \theta(y)$. $\qquad\square$

The construction of the lexicographic centre enables us to compute the nucleolus of a game through a sequence of linear programs. Consider first (4.129)–(4.131) for $k = 1$. Equation (4.129) gives

$$\varepsilon^1 = \min_{x \in X^0} \max_{S \in \Sigma^0} e(S, x).$$

Let

$$\max_{S \in \Sigma^0} e(S, x) = \alpha.$$

Then

$$\alpha \geq e(S, x) = v(S) - x(S), \qquad S \in \Sigma^0,$$

or

$$x(S) + \alpha \geqslant v(S), \qquad S \in \Sigma^0.$$

The problem of determining ε^1, X^1 and Σ_1 is equivalent to the solution of the following linear program:

Minimize α,
subject to
$$x(S) + \alpha \geqslant v(S), \qquad S \in \Sigma^0, \tag{4.137}$$
$$x \in X^0.$$

The minimum of this linear program is ε^1, which is the minimum of the maximum excess and is attained over a set X^1 of imputations through a set Σ_1 of coalitions, i.e. for all $S \in \Sigma_1$ and all $x \in X^1$ we have $e(S, x) = \varepsilon^1$.

Next we put aside the coalitions of Σ_1 and consider the linear program:

Minimize α,
subject to
$$x(S) + \alpha \geqslant v(S), \qquad S \in \Sigma^1 = \Sigma^0 \backslash \Sigma_1, \tag{4.138}$$
$$x \in X^1.$$

This linear program will yield the second-largest excess ε^2 as well as the corresponding X^2 of (4.130) and Σ_2 of (4.131). The coalitions of Σ_2 are now put aside and the procedure is repeated until there are no coalitions left. We eventually obtain a unique imputation which is the nucleolus of the game.

Example 4.14. Let us consider once again the three-person game Γ of Example 4.11 in the preceding section. The characteristic function v has the values:

$$v(i) = 0, \qquad i = 1, 2, 3,$$
$$v(12) = \tfrac{1}{3}, \qquad v(13) = \tfrac{1}{6}, \qquad v(23) = \tfrac{5}{6},$$
$$v(123) = 1.$$

The first linear program (4.137) gives

$$\alpha = \varepsilon^1 = -\tfrac{1}{12},$$
$$X^1 = \{x : x_1 = \tfrac{1}{12}, x_2 \leqslant \tfrac{9}{12}, x_3 \leqslant \tfrac{7}{12}, x_2 + x_3 = \tfrac{11}{12}\},$$
$$\Sigma_1 = \{\{1\}, \{2, 3\}\},$$
$$\Sigma^1 = \Sigma^0 \backslash \Sigma_1 = \{\{2\}, \{3\}, \{1, 2\}, \{1, 3\}\}.$$

The set X^1 is the least core of the game.

The second linear program (4.138) yields

$$\alpha = \varepsilon^2 = -\tfrac{7}{24},$$
$$X^2 = \{x\colon x_1 = \tfrac{1}{12}, x_2 = \tfrac{13}{24}, x_3 = \tfrac{9}{24}\},$$
$$\Sigma_2 = \{\{1, 2\}, \{1, 3\}\},$$
$$\Sigma^2 = \Sigma^1\backslash\Sigma_2 = \{\{2\}, \{3\}\}.$$

Similarly, the third linear program gives

$$\alpha = \varepsilon^3 = -\tfrac{9}{24},$$
$$X^3 = X^2,$$
$$\Sigma_3 = \{\{3\}\},$$
$$\Sigma^3 = \Sigma^2\backslash\Sigma_3 = \{\{2\}\};$$

and the fourth and final linear program gives

$$\alpha = \varepsilon^4 = -\tfrac{13}{24},$$
$$X^4 = X^3 = X^2,$$
$$\Sigma_4 = \{\{2\}\},$$
$$\Sigma^4 = \Sigma^3\backslash\Sigma_4 = \varnothing.$$

Hence we have

$$N(\Gamma) = X^4 = \{(\tfrac{1}{12}, \tfrac{13}{24}, \tfrac{9}{24})\}.$$

Example 4.15. The characteristic function v of a four-person cooperative game Γ has the values:

$$v(i) = 0, \qquad i = 1, 2, 3, 4,$$
$$v(12) = v(34) = v(14) = v(23) = 1,$$
$$v(13) = \tfrac{1}{2}, \quad v(24) = 0,$$
$$v(123) = v(124) = v(134) = v(234) = 1,$$
$$v(1234) = 2.$$

The nucleolus of this game is obtained by solving a sequence of three linear programs. It is easily verified that starting from $X^0 = X$ and $\Sigma^0 = \{S\colon S \subset I; S \neq \varnothing, I\}$ we have:

(i) $\varepsilon^1 = 0,$
$$X^1 = \{(1 - \tfrac{3}{4}\lambda, \tfrac{3}{4}\lambda, 1 - \tfrac{3}{4}\lambda, \tfrac{3}{4}\lambda)\}, \qquad 0 \leq \lambda \leq 1,$$
$$\Sigma_1 = \{\{1, 2\}, \{3, 4\}, \{1, 4\}, \{2, 3\}\},$$
$$\Sigma^1 = \Sigma^0\backslash\Sigma_1 = \{\{1, 2, 3\}, \{1, 2, 4\}, \{1, 3, 4\}, \{2, 3, 4\},$$
$$\{1, 3\}, \{2, 4\}, \{1\}, \{2\}, \{3\}, \{4\}\};$$

(ii) $\varepsilon^2 = -\frac{1}{2}$,

$X^2 = \{(\frac{1}{2}, \frac{1}{2}, \frac{1}{2}, \frac{1}{2})\}$,

$\Sigma_2 = \{\{1, 2, 3\}, \{1, 2, 4\}, \{1, 3, 4\}, \{2, 3, 4\},$
$\qquad \{1, 3\}, \{1\}, \{2\}, \{3\}, \{4\}\}$,

$\Sigma^2 = \Sigma^1 \backslash \Sigma_2 = \{\{2, 4\}\}$;

(iii) $\varepsilon^3 = -1$.

$X^3 = X^2 = \{(\frac{1}{2}, \frac{1}{2}, \frac{1}{2}, \frac{1}{2})\}$,

$\Sigma_3 = \{\{2, 4\}\}$,

$\Sigma^3 = \Sigma^2 \backslash \Sigma_3 = \varnothing$.

Hence the nucleolus of the game is

$$N(\Gamma) = X^3 = \{(\tfrac{1}{2}, \tfrac{1}{2}, \tfrac{1}{2}, \tfrac{1}{2})\}.$$

4.12 Shapley value

Shapley value is another solution concept for the cooperative game. As early as 1952, L. S. Shapley discovered the well-known formula bearing his name. Shapley (1953) proves that three axioms determine a unique set of values which can be treated as a distribution scheme for the players of a cooperative game.

As before, let $\Gamma \equiv [I, v]$ be an *n*-person cooperative game whose characteristic function v satifies the conditions

$$v(\varnothing) = 0, \qquad v(I) \geqslant \sum_{i=1}^{n} v(i).$$

Since an *n*-person cooperative game $\Gamma \equiv [I, v]$ is completely determined by its characteristic function v, we shall also refer to v as the game.

Definition 4.18. Let $N \subseteq I$ be a coalition. If

$$v(S) = v(S \cap N) + \sum_{i \in S \backslash N} v(i) \tag{4.139}$$

for all $S \subseteq I$, we say that N is a *carrier* of the game Γ.

Any superset of a carrier of a game v is again a carrier of v.

Proof. Let N be a carrier of v. Then

$$v(S) = v(S \cap N) + \sum_{i \in S \backslash N} v(i)$$

for every $S \subseteq I$. Let $N' \supset N$. Then

$$v(S \cap N') = v((S \cap N') \cap N) + \sum_{i \in (S \cap N') \backslash N} v(i)$$

$$= v(S \cap N) + \sum_{i \in (S \cap N') \backslash N} v(i)$$

$$= v(S) - \sum_{i \in S \backslash N} v(i) + \sum_{i \in (S \cap N') \backslash N} v(i).$$

We have, for every $S \subseteq I$,

$$v(S) = v(S \cap N') + \sum_{i \in S \setminus N} v(i) - \sum_{i \in (S \cap N') \setminus N} v(i)$$

$$= v(S \cap N') + \sum_{i \in (S \setminus (S \cap N')) \setminus N} v(i)$$

$$= v(S \cap N') + \sum_{i \in S \setminus N'} v(i).$$

Hence N' is also a carrier of v.

If a player does not belong to some carrier, then this player cannot make any extra contribution to any coalition into which he is allowed to enter.

Let π be a permutation of $I = \{1, \ldots, n\}$, i.e. a one-to-one mapping of I onto itself. πS is the image of the coalition S under π. If $\pi i = j$, then j is the image of i under π. Denote by $\pi(I)$ the set of all permutations of I.

Definition 4.19. The *Shapley value* of an n-person cooperative game $\Gamma \equiv [I, v]$ is the function

$$\Phi(v) = (\Phi_1(v), \ldots, \Phi_n(v))$$

satisfying the following axioms.

Axiom 1. Axiom of symmetry. If for every $\pi \in \pi(I)$ and every $S \subseteq I$ we have

$$v(\pi S) = v(S),$$

then

$$\Phi_{\pi i}(v) = \Phi_i(v).$$

Axiom 2. Axiom of efficiency. For every carrier N of Γ,

$$\sum_{i \in N} \Phi_i(v) = v(N).$$

Axiom 3. Law of aggregation. For any two characteristic functions v and w defined on the set of all subsets of I,

$$\Phi(v + w) = \Phi(v) + \Phi(w).$$

Lemma 1. *If N is a carrier of $\Gamma \equiv [I, v]$, then, for $i \notin N$,*

$$\Phi_i(v) = v(i).$$

Proof. Suppose $i \notin N$. Since N is a carrier of Γ, we have

$$v(N \cup i) = v([N \cup i] \cap N) + \sum_{k \in (N \cup i) \setminus N} v(k) = v(N) + v(i).$$

$N \cup i$ is also a carrier. Hence, by Axiom 2,

$$\sum_{k \in N \cup i} \Phi_k(v) = v(N) + \Phi_i(v) = v(N \cup i).$$

Thus
$$\Phi_i(v) = v(i). \qquad \square$$

Consider the following game. Let $R \subseteq I$, $R \neq \emptyset$. For every $S \subseteq I$, define

$$v_R(S) = \begin{cases} 1, & \text{if } S \supseteq R, \\ 0, & \text{if } S \not\supseteq R. \end{cases} \qquad (4.140)$$

It is easily verified that, for any $c > 0$, cv_R is a characteristic function, and R is a carrier.

Lemma 2. *Let* $c > 0$, $R \subseteq I$, $|R| > 0$. *Then*

$$\Phi_i(cv_R) = \begin{cases} c/|R|, & \text{if } i \in R, \\ 0, & \text{if } i \notin R. \end{cases} \qquad (4.141)$$

Proof. Let $i, j \in R$. Choose a permutation $\pi \in \pi(I)$ such that

$$\pi R = R, \qquad \pi i = j,$$

then $i \notin R$ implies $\pi i \notin R$. We have

$$cv_R(S) = \begin{cases} c, & \text{if } R \supseteq R, \\ 0, & \text{if } S \not\supseteq R, \end{cases}$$

and R is a carrier of cv_R.
 By Axiom 1,

$$\Phi_j(cv_R) = \Phi_i(cv_R), \qquad i, j \in R.$$

By Axiom 2,

$$|R| \, \Phi_i(cv_R) = \sum_{i \in R} \Phi_i(cv_R) = cv_R(R) = c.$$

Hence
$$\Phi_i(cv_R) = c/|R|, \qquad i \in R.$$

If $i \notin R$, by Lemma 1,

$$\Phi_i(cv_R) = cv_R(i) = 0. \qquad \square$$

Lemma 3. *The characteristic function of an n-person cooperative game* $\Gamma \equiv [I, v]$ *can be expressed as a linear combination of* v_R:

$$v = \sum_{R \subseteq I} \lambda_R v_R, \qquad (4.142)$$

where the coefficients λ_R are given by

$$\lambda_R = \sum_{T \subseteq R} (-1)^{|R|-|T|} v(T). \tag{4.143}$$

Proof. Substituting (4.143) into (4.142) we obtain

$$v(S) = \sum_{R \subseteq I} \left[\sum_{T \subseteq R} (-1)^{|R|-|T|} v(T) \right] v_R(S) \tag{4.144}$$

for every $S \subseteq I$. In the above expression, $v_R(S)$ vanishes except for $R \subseteq S$. When $R \subseteq S$, $v_R(S) = 1$ by (4.140). Therefore,

$$v(S) = \sum_{R \subseteq S} \left[\sum_{T \subseteq R} (-1)^{|R|-|T|} v(T) \right].$$

Exchanging the order of summations gives

$$v(S) = \sum_{T \subseteq S} \sum_{\substack{R \subseteq S \\ R \supseteq T}} (-1)^{|R|-|T|} v(T)$$

$$= \sum_{T \subseteq S} \left[\sum_{|R|=|T|}^{|S|} (-1)^{|R|-|T|} \binom{|S|-|T|}{|R|-|T|} \right] v(T),$$

where

$$\binom{|S|-|T|}{|R|-|T|} = C_{|R|-|T|}^{|S|-|T|}$$

is the notation of combination. The expression in brackets vanishes except for $|T| = |S|$. When $|T| = |S|$, its value is 1. Thus

$$v(S) = v(S) + \sum_{\substack{T \subseteq S \\ T \neq S}} (1-1)^{|S|-|T|} v(T) = v(S).$$

This proves that (4.144) holds for $S \subseteq I$. □

Now let us evaluate $\Phi_i(v)$. By (4.142),

$$v = \sum_{R \subseteq I} \lambda_R v_R$$

$$= \sum_{\substack{R \subseteq I \\ \lambda_R \geqslant 0}} \lambda_R v_R - \sum_{\substack{R \subseteq I \\ \lambda_R < 0}} (-\lambda_R) v_R.$$

Using Axiom 3 we obtain

$$\Phi_i(v) = \sum_{\substack{R \subseteq I \\ \lambda_R \geqslant 0}} \Phi_i(\lambda_R v_R) - \sum_{\substack{R \subseteq I \\ \lambda_R < 0}} \Phi_i((-\lambda_R) v_R).$$

By (4.141), $\Phi_i(cv_R) = 0$ if $i \notin R$. Hence the above equation reduces to

$$\Phi_i(v) = \sum_{\substack{i \in R \subseteq I \\ \lambda_R \geqslant 0}} \frac{\lambda_R}{|R|} + \sum_{\substack{i \in R \subseteq I \\ \lambda_R < 0}} \frac{\lambda_R}{|R|},$$

or

$$\Phi_i(v) = \sum_{\substack{R \subseteq I \\ R \ni i}} \frac{\lambda_R}{|R|}. \tag{4.145}$$

Substituting (4.143) into (4.145) gives

$$\Phi_i(v) = \sum_{\substack{R \subseteq I \\ R \ni i}} \frac{1}{|R|} \sum_{S \subseteq R} (-1)^{|R|-|S|} v(S)$$

$$= \sum_{S \subseteq I} \left[\sum_{\substack{R \supseteq S \\ R \ni i}} \frac{1}{|R|} (-1)^{|R|-|S|} \right] v(S),$$

or

$$\Phi_i(v) = \sum_{\substack{S \subseteq I \\ S \ni i}} \left[\sum_{R \supseteq S} \frac{1}{|R|} (-1)^{|R|-|S|} \right] v(S)$$

$$+ \sum_{\substack{S \subseteq I \\ S \not\ni i}} \left[\sum_{R \supseteq S \cup i} \frac{1}{|R|} (-1)^{|R|-|S|} \right] v(S). \tag{4.146}$$

We have

$$\sum_{R \supseteq S} \frac{1}{|R|} (-1)^{|R|-|S|} = \sum_{|R|=|S|}^{n} \frac{1}{|R|} (-1)^{|R|-|S|} \binom{n-|S|}{|R|-|S|}$$

$$= \frac{(n-|S|)!(|S|-1)!}{n!}. \tag{4.147}$$

In the last term in (4.146), denote $S \cup i$ by S', then $S = S' \backslash i$, and that term reduces to

$$\sum_{\substack{S' \backslash i \subseteq I \\ S' \ni i}} \sum_{R \supseteq S'} \frac{1}{|R|} (-1)^{|R|-|S'|+1} v(S' \backslash i)$$

$$= - \sum_{\substack{S' \subseteq I \\ S' \ni i}} \sum_{R \supseteq S'} \frac{1}{|R|} (-1)^{|R|-|S'|} v(S' \backslash i)$$

$$= - \sum_{\substack{S \subseteq I \\ S \ni i}} \sum_{R \supseteq S} \frac{1}{|R|} (-1)^{|R|-|S|} v(S \backslash i). \tag{4.148}$$

Substituting (4.147) and (4.148) back into (4.146), we obtain

$$\Phi_i(v) = \sum_{S \ni i} \frac{(n - |S|)!(|S| - 1)!}{n!} [v(S) - v(S\backslash i)]. \qquad (4.149)$$

The Shapley value of the game v is obtained by letting $i = 1, \ldots, n$ in the above expression:

$$\Phi(v) = (\Phi_1(v), \ldots, \Phi_n(v)).$$

Theorem 4.15. *Let* $\Gamma \equiv [I, v]$ *be an* n-*person cooperative game. Then there exists a unique Shapley value* $\Phi(v)$ *given by*

$$\Phi_i(v) = \sum_{S \ni i} \frac{(n - |S|)!(|S| - 1)!}{n!} [v(S) - v(S\backslash i)], \qquad i = 1, \ldots, n. \quad (4.149)$$

Proof. We should verify that Φ satisfies the three axioms in Definition 4.19.

(1) *Axiom of Symmetry.* Let $\pi \in \pi(I)$, $v(\pi S) = v(S)$. Then $|\pi S| = |S|$. By (4.149),

$$\Phi_{\pi i}(v) = \sum_{\pi S \subseteq \pi I} \frac{(n - |\pi S|)!(|\pi S| - 1)!}{n!} [v(\pi S) - v(\pi(S\backslash i))]$$

$$= \sum_{S \subseteq I} \frac{(n - |S|)!(|S| - 1)!}{n!} [v(S) - v(S\backslash i)]$$

$$= \Phi_i(v).$$

(2) *Axiom of efficiency.* If i does not belong to some carrier, it does not belong to the minimal carrier (i.e. the intersection of all carriers). Hence, for every $S \subseteq I$, $S \ni i$, we have

$$v(S) - v(S\backslash i) = v(i).$$

Substituting into (4.149) gives

$$\Phi_i(v) = v(i) \sum_{\substack{S \subseteq I \\ S \ni i}} \frac{(n - |S|)!(|S| - 1)!}{n!}.$$

In the sum on the right-hand side of the above equation, we combine those terms which correspond to the S with the same number of players. For example, the number of coalitions S with k players including i is

$$\binom{n - 1}{k - 1}.$$

Hence

$$\Phi_i(v) = v(i) \sum_{k=1}^{n} \sum_{|S|=k} \frac{(n-|S|)!(|S|-1)!}{n!}$$

$$= v(i) \sum_{k=1}^{n} \frac{(n-k)!(k-1)!}{n!} \binom{n-1}{k-1}$$

$$\overset{.}{=} v(i) \sum_{k=1}^{n} \frac{1}{n} = v(i) \cdot n \cdot \frac{1}{n} = v(i). \qquad (4.150)$$

We now proceed to prove

$$\sum_{i \in I} \Phi_i(v) = v(I).$$

By (4.149),

$$\sum_{i \in I} \Phi_i(v) = \sum_{i=1}^{n} \sum_{S \ni i} \frac{(n-|S|)!(|S|-1)!}{n!} [v(S) - v(S \backslash i)]. \qquad (4.151)$$

Consider a fixed coalition $R \subseteq I$. In the double sum on the right-hand side of the last equation, $v(R)$ appears in the brackets as a minuend $|R|$ times, and thus the sum of coefficients corresponding to these $v(R)$ is

$$|R| \frac{(n-|R|)!(|R|-1)!}{n!} = \frac{(n-|R|)!|R|!}{n!}. \qquad (4.152)$$

This equation holds for all $R \subseteq I$. When $R = I$, the expression in (4.152) equals 1. That is to say, the coefficient of $v(I)$ as a minuend in the brackets is 1.

The second term $v(S \backslash i)$ in the brackets equals $v(R)$ for $n - |R|$ times, since i can be any one element of $I \backslash R$. Hence the sum of coefficients corresponding to these $v(R)$ in the double sum is

$$-(n-|R|) \frac{[n-(|R|+1)]!(|R|+1-1)!}{n!} = -\frac{(n-|R|)!|R|!}{n!}. \qquad (4.153)$$

This equation holds for all $|R| \neq n$, i.e. for all $R \neq I$. When $R = I$, the expression in (4.153) vanishes.

Thus we see that in the right-hand side of (4.151) there remains one term only, i.e. $v(I)$, whose coefficient is 1. Other terms offset one another. Hence we have

$$\sum_{i \in I} \Phi_i(v) = v(I). \qquad (4.154)$$

Now suppose N is a carrier of the game v. By definition (4.139),

$$v(I) = v(I \cap N) + \sum_{i \in I \backslash N} v(i),$$

i.e.

$$v(I) = v(N) + \sum_{i \notin N} v(i).$$

By (4.154),

$$v(I) = \sum_{i \in I} \Phi_i(v);$$

and by (4.150),

$$\sum_{i \notin N} v(i) = \sum_{i \notin N} \Phi_i(v).$$

Therefore,

$$v(N) = \sum_{i \in I} \Phi_i(v) - \sum_{i \notin N} \Phi_i(v) = \sum_{i \in N} \Phi_i(v).$$

This completes the proof that Φ satisfies the axiom of efficiency.

(3) *Law of aggregation.* The law of aggregation is satisfied since each Φ_i is a linear combination of the values of v.

That the Shapley value (4.149) is unique can be seen directly from its derivation. $\qquad\square$

Example 4.16. The characteristic function v of the three-person coopera-tive game of Example 4.11 has the values:

$$v(i) = 0, \qquad i = 1, 2, 3,$$
$$v(12) = \tfrac{1}{3}, \qquad v(13) = \tfrac{1}{6}, \qquad v(23) = \tfrac{5}{6},$$
$$v(123) = 1.$$

We know from Sections 4.9 and 4.10 that the least core of the game is

$$LC(\Gamma) = C_{-\frac{1}{12}}(\Gamma) = \{x : x_1 = \tfrac{1}{12}, x_2 \leqslant \tfrac{9}{12}, x_3 \leqslant \tfrac{7}{12}\},$$

and that the pre-kernel and kernel are the mid-point of the least core:

$$K^*(\Gamma) = K(\Gamma) = \{(\tfrac{1}{12}, \tfrac{13}{24}, \tfrac{9}{24})\},$$

which is also the nucleolus of the game.

The Shapley value of this game is

$$\Phi(v) = (\tfrac{5}{36}, \tfrac{17}{36}, \tfrac{14}{36}).$$

Example 4.17. An example often referred to in the literature of game theory is the so called 'voting game'. Suppose there are five players. Player 1 has three votes, players 2, 3, 4, 5 has one vote each. A coalition with four or more votes wins. We shall represent a winning coalition by assigning the value 1 to the characteristic function v and a losing coalition by the value 0. Thus the characteristic function v of this game Γ is

$$v(\varnothing) = 0,$$
$$v(12) = v(13) = v(14) = v(15) = 1,$$
$$v(123) = v(124) = v(125) = v(134) = v(135) = v(145) = 1,$$
$$v(1234) = v(1235) = v(1245) = v(1345) = v(2345) = 1,$$
$$v(12345) = 1, \qquad v(S) = 0 \quad \text{for all other } S.$$

The Shapley value is easily calculated:

$$\Phi(v) = (\tfrac{6}{10}, \tfrac{1}{10}, \tfrac{1}{10}, \tfrac{1}{10}, \tfrac{1}{10}).$$

The kernel $K(\Gamma)$ and nucleolus $N(\Gamma)$ of this game with respect to the grand coalition are the same:

$$K(\Gamma) = N(\Gamma) = \{(\tfrac{3}{7}, \tfrac{1}{7}, \tfrac{1}{7}, \tfrac{1}{7}, \tfrac{1}{7})\}$$

(cf. Davis and Maschler, 1965).

REFERENCES

Aumann, R. J., Peleg, B., and Rabinowitz, P. (1965). A method for computing the kernel of n-person games. *Mathematics of Computation,* **19,** 531–51.

Bruyneel, G. (1978). On balanced sets, with applications in game theory. *Bulletin de la Société Mathématique de Belgique,* **30,** 93–108.

Dantzig, G. B. (1951). A Proof of the equivalence of the programming problem and the game problem. In *Activity Analysis of Production and Allocation,* Cowles Commission Monograph No. 13, pp. 330–35. Wiley, New York.

Dantzig, G. B. (1956). Constructive proof of the min-max theorem. *Pacific Journal of Mathematics,* **6,** 25–33.

Davis, M. and Maschler, M. (1965). The kernel of a cooperative game. *Naval Research Logistics Quarterly,* **12,** 223–59.

Dragan, I. (1981). A procedure for finding the nucleolus of a coopeative n-person game. *Zeitschrift für Operations Research,* **25,** 119–31.

Dresher, M. (1961). *Games of Strategy, Theory and Applications.* Prentice-Hall, Englewood Cliffs, N.J.

Glicksman, A. M. (1963). *An Introduction to Linear Programming and the Theory of Games.* Wiley, New York.

Jones, A. J. (1980). *Game Theory: Mathematical Models of Conflict.* Ellis Horwood, Chichester.

Karlin, Samuel (1959). *Mathematical Methods and Theory in Games, Programming and Economics.* Addison-Wesley, Reading, Mass.

Kohlberg, E. (1971). On the nucleolus of a characteristic function game. *SIAM Journal of Applied Mathematics,* **20**(1), 62–6.

Kohlberg, E. (1972). The nucleolus as a solution of a minimization problem. *SIAM Journal of Applied Mathematics,* **23**(1), 34–9.

Loomis, L. H. (1946). On a theorem of von Neumann. *Proceedings of the National Academy of Sciences of the U.S.A.,* **32,** 213–5.

Lucas, W. F. (1969). The proof that a game may not have a solution. *Transactions of the American Mathematical Society,* **137** (402), 219–29.

Lucas, W. F. (1971). Some recent developments in n-person game theory. *SIAM Review,* **13**(4), 491–523.

Lucas, W. F. and Rabie, M. (1982). *Games with no solutions and empty cores, Mathematics of Operations Research,* **7**(4), 491–500.

Luce, R. D. and Raiffa, H. (1957). *Games and Decisions.* Wiley, New York.

McKinsey, J. C. C. (1952). *Introduction to the Theory of Games.* McGraw-Hill, New York.

Maschler, M. and Peleg, B. (1966). A characterization, existence proof and dimension bounds for the kernel of a game. *Pacific Journal of Mathematics,* 289–328.

Maschler, M., Peleg, B., and Shapley, L. S. (1972). The kernel and bargaining set for convex games. *International Journal of Game Theory,* **1,** 73–9.

Maschler, M., Peleg, B., and Shapley, L. S. (1979). Geometric properties of the kernel, nucleolus, and related solution concepts. *Mathematics of Operations Research*, **4**(4), 303–38.

Nash, J. F. (1950). Equilibrium points in *n*-person games. *Proceedings of the National Academy of Sciences of the U.S.A.*, **36**, 48–9.

Nash, J. F. (1951). Non-cooperative games. *Annals of Mathematics*, **54**(2), 286–95.

von Neumann, J. (1928). Zur Theorie der Gesellschaftsspiele. *Mathematische Annalen*, **100**, 295–320.

von Neumann, J. (1937). Über ein ökonomisches Gleichungssystem und eine Verallgemeinerung des Brouwerschen Fixpunktsatzes. *Ergebnisse eines Mathematische Kolloquiums*, **8**, 73–83.

von Neumann, J. (1959). On the theory of games of strategy. In *Contributions to the Theory of Games,* Vol. IV, Annals of Mathematics Studies 40, pp. 13–42. (English translation of von Neumann (1928).)

von Neumann, J. and Morgenstern, O. (1953). *Theory of Games and Economic Behavior* (3rd edn). Princeton University Press, Princeton, New Jersey.

Owen, G. (1967). An elementary proof of the minimax theorem. *Management Science*, **13**(9), 765.

Owen, G. (1982). *Game Theory* (2nd edn). Academic Press, New York.

Peleg, B. (1966). The kernel of the general-sum four-person game. *Canadian Journal of Mathematics*, **18**(4), 673–7.

Rosenmüller, J. (1981). *The Theory of Games and Markets*. North-Holland, Amsterdam.

Schmeidler, D. (1969). The nucleolus of a characteristic function game. *SIAM Journal of Applied Mathematics*, **17**(6), 1163–70.

Shapley, L. S. (1953). A value for *n*-person games. In *Contributions to the Theory of Games,* Vol. II, Annals of Mathematics Studies 28 (ed. Kuhn and Tucker), pp. 307–17.

Vajda, S. (1956). *The Theory of Games and Linear Programming*. Methuen, London.

Ville, J. A. (1938). Sur la Théorie Génerale des jeux ou intervient l'habilité des joueurs, Applications aux Jeux de Hasard, by Émile Borel and Jean Ville, Tom IV, Fascicule II. In *Traité du Calcul des Probabilités et de ses Applications* par Émile Borel, pp. 105–13. Gauthier-Villars, Paris.

Vorob'ev, N. N. (1977). *Game Theory, Lectures for Economists and Systems Scientists*. Springer-Verlag, New York.

Wang, Jianhua (1982). An inductive proof of von Neumann's minimax theorem. *Chinese Journal of Operations Research*, **1**(1), 68–70.

Wang, Jianhua (1983). Inductive proof of a saddle point theorem. *Journal of Mathematical Research and Exposition*, **3**(1), 142–4.

Wang, Jianhua (1986a). Cooperative games. *Chinese Journal of Operations Research*, **5**(2), 1–9.

Wang, Jianhua (1986b). *The Theory of Games* (in Chinese). Tsinghua University Press, Beijing.

Williams, J. D. (1954). *The Compleat Strategyst*. McGraw-Hill, New York.

INDEX

INDEX